1001

IDEAS

for Science Projects
on the Environment

ARCO

1001 IDEAS

for Science Projects
on the Environment

Marion A. Brisk, Ph.D.
Center for Biomedical Education at
City College of New York

Macmillan • USA

Macmillan General Reference USA
A Simon & Schuster Macmillan Company
1633 Broadway
New York, NY 10019-6785

An Arco Book

ARCO and colophon is a registered trademark of Simon & Schuster Inc.
MACMILLAN and colophon is a registered trademark of Macmillan,
Inc.

Manufactured in the United States of America
10 9 8 7 6 5 4 3 2 1

Library of Congress Number: 97-070061
ISBN: 0-02-861707-X

Interior and cover design by Scott Meola

507.8
B

CONTENTS

PART III
Resources

To the Reader

You can use this book for several purposes. Its most obvious purpose is to help you discover a project on the environment that you will find interesting and that you will be able to complete successfully. You can use the book to gain an overview of the major environmental problems of our time and their known causes and to learn about specific topics of the environment that are of special interest to you. You can also use this book as a guide to environmental resources, including many of those on the Internet.

The first part of the book, "Helpful Hints," contains suggestions on various aspects of your project. It begins by addressing "How to Select a Topic on the Environment," providing you with suggestions on how to choose a project on the environment that you will enjoy and also be able to successfully complete. In this part of the book you will find a description of our system of symbols. These symbols give you information about important aspects of the topic to help you select one—the level of difficulty, for example, or the need for special equipment or supplies. "How to Find Information About the Environment" provides you with suggestions on how to go about researching the environmental issues addressed by your project or report along with how to tap into the extensive information on the Internet. "How to Write a Research Paper on the Environment" follows with guidelines on presenting your work according to scientific methods and procedures. Because many environmental problems have strong links to political, economic, and social issues, it is especially critical that scientific approaches be used to study the impact of human activities on the air, water, land, and living world.

The second part of this book, "1001 Ideas for Projects, Papers, and Reports on the Environment," presents hundreds of important and timely project ideas. These ideas are separated into six categories: air, land, water, the living world, energy, and human issues. Among these you will find many different ideas about a variety of important environmental issues. With the numerous ideas presented and their variety, you should be able to locate a topic that intrigues you. Each of the six categories begins with general information followed by an overview of the major environmental issues concerning that category. A brief description and some background information are provided along with each of the many ideas that follow. In addition, to get you started, each topic offers several places to begin your research—journal articles, books, agencies, organizations, or Web sites on the World Wide Web (WWW). The number of environmental sites on the WWW has increased enormously in the last few years and continues to grow at an astounding rate. The WWW in particular offers volumes of information about most environmental issues. Remember: All scientific studies begin with a literature search.

Regardless of which topic you select, it will be helpful if you also read the introduction to the main category. If you choose a topic involving air—say, global warming, for example—it will prove useful for you to read the introduction to the category "Air" in order to gain an overview of the problem before you begin. In fact, just by reading these introductions you will gain some knowledge of critical environmental concerns.

The last section of this book is a very important one. Part III, "Resources," lists the names and addresses of the environmental organizations and agencies that are mentioned in the ideas section along with their Web sites where applicable. Part III also includes a list of Web sites that have information on many environmental issues. As you visit these Web sites, you are likely to find links to your specific topic of interest.

PART I

HELPFUL HINTS

To the Student

If you are like most students, you will find doing a report or project on the environment to be very interesting, stimulating, and relevant to your life. In this section are some suggestions that will help you successfully complete your assignment as well as enjoy your topic. A very important part of your work will involve researching your idea; many environmental issues are constantly changing so you must make sure your sources are current. At the end of this section, you will find some general environmental sources as well as Web sites to help you get started. The World Wide Web in particular has become an excellent source of environmental information and therefore relevant Web sites have been included for each idea.

How to Select a Topic on the Environment

It is most important that you select a topic that you not only like, but also one that you can complete within the time-frame of your assignment and within the resources available to you. Following are some suggestions on how to choose a suitable topic:

1. Select an idea that is particularly interesting and relevant to you.

Are you fascinated by ocean fish? Do you enjoy hiking in forests? Does air pollution concern you? All of us are affected by many environmental issues. Which current topics on the environment do you tend to listen to on news programs or read about in newspapers? Are there pressing problems with the environment in your own community? Try to focus on an idea that appeals to you and is also particularly important to you. If you enjoy feeding and observing birds, you might be interested in studying the causes of their declining numbers. If you like gardening, you might enjoy a topic concerning the health of agricultural soil or sustainable farming. By selecting a topic that you really like, you will be more likely to remain motivated as you research the literature and consult agencies and organizations for current information. It is also a good idea to work on a project that will be of interest to your reader(s). All of the ideas in this book concern contemporary environmental issues that tend to be of general interest.

2. Select an idea that you will have time to complete.

An assignment that is due at the end of the year clearly is expected to be more comprehensive than one that is completed after one semester. Therefore, make sure that your idea meets the requirements of the assignment and that it can be completed in the allotted time. Will you have enough time to gather information and the necessary materials? How often can you work on your project? Will you have time each day or only on weekends? It might be helpful to record your schedule for a week to see what time you actually have available for your project. Some projects require large blocks of time which you may not have available. It may be helpful to do some general reading on the subject to help ascertain the time demands of an idea before you actually commit yourself to any one project.

Try to be as clear as possible about what you plan to do so that you do not get sidetracked along the way. It may be that as you progress some changes may be in order. Try again to clearly define your new plans and new time requirements if any. Be aware that researching environmental topics tends to be very time-consuming because of the need to find the most current information. You may need to use various state or federal agencies or scientific organizations to help you find the relevant facts. If you use the Internet or World Wide Web, finding relevant information will be exciting but most likely also time-consuming.

3. Determine what special equipment, facilities, or resources you will need.

Plan out your project as specifically as possible in order to identify what you will need along the way. Do you have access to the facilities and equipment required? Will you be able to gather sufficient information in the allotted time? Can you use a library that contains the periodicals that you need? Because information on environmental topics is constantly changing, the resources you rely on are particularly important. Before you decide on a topic, it is generally a good idea to do some research first to ascertain what you will need in order to successfully complete it.

You may also find that you will need to make changes in the project as you progress. Be careful about any equipment or supplies you bring into your home, especially if young children or animals are part of your family. Before using any chemical, make sure you are familiar with its properties and how it can be handled and discarded safely. Look it up in the Merck Index. This reference is generally available in university libraries and in most general libraries as well.

Discuss your ideas with teachers and other professionals to get their advice and suggestions. There are often environmental professionals who will answer your questions at federal, state, and local agencies and organizations.

4. Consider the level of difficulty of your topic.

Make sure that you have sufficient background knowledge in the area you have selected so that you will be able to understand the literature and be able to write the report or complete the project. If you choose a topic that becomes too technical, you may end up spending way too much time acquiring the required background information and then run out of time for the project. To lessen the risk of this occurring, as already mentioned, briefly research the topic in the beginning. If you are having difficulty understanding the literature, select another topic or make your topic more general and less technical.

To help you select a suitable topic, a set of symbols appears next to each idea. These symbols and their meanings are listed below:

●   This topic is of average difficulty and is suitable for most high school and undergraduate assignments.

△   This idea involves a significant time and work commitment and is thus suitable for a long-term project.

This project requires special facilities, equipment, or supplies.

○ A large public or college library with extensive resources will be required to adequately research this topic.

* This project involves supplies, equipment, or techniques that could be harmful. Make sure you have adequate direction on how to safely use all supplies and equipment and how to perform all required procedures.

✚ This idea involves very technical information and requires previous knowledge.

How to Find Information About the Environment

There are numerous sources of environmental information available today. They include libraries, of course; government agencies and organizations; professional, scientific, and environmental organizations; and the Internet and the World Wide Web. These sources are briefly described below, and some specific sources are provided in Part III of this book. The World Wide Web, in particular, offers a wealth of information on a wide variety of environmental topics. Web sites are provided, where possible, for the individual ideas; some additional ones are given in Part III. Note that each Web site will have links to many others so that you will have access to a large quantity of information. There are also directories of environmental sources that you can consult. One of these is:

> *Directory of Environmental Information Sources,*
> 5th edition, 1995
> Edited by Thomas F.P. Sultan
> 4 Research Place, Suite 200
> Rockville, MD 20850

USING THE LIBRARY

Once you have selected an idea for either a paper or a project, begin your research by consulting general reference sources like dictionaries, handbooks, and encyclopedias. These general sources will provide you with some basic information. For example, if your topic is about soil loss, you will need to know about the composition of soil, how it forms, and how it naturally erodes. Only when you understand the natural processes can you fully comprehend how human activities are interfering and causing environmental damage. If you are researching pollution of a river, it will be necessary to understand how streams and rivers receive water from the runoff of the entire watershed region. Some general sources include *Van Nostrand's Scientific Encyclopedia,* the *Merck Manual,* the *McGraw-Hill Encyclopedia of Science and Technology,* and the *McGraw-Hill Encyclopedia of Environmental Science and Engineering.* You can also consult one of the many environmental science textbooks that are available. Look for them in a university bookstore or at the library. You can locate general sources in most libraries by using the library card or microfiche catalogs or by doing a computer search of the library's holdings. It's a good idea to take notes throughout your reading, recording important and relevant information and accurately recording each reference in case you need to consult it again or later must cite it in your report.

After you have acquired some general knowledge about your topic, search for recent major articles addressing the relevant environmental issues. Many of these important articles have appeared in the journals *Science, Nature, Environmental Science and Technology,* and *Scientific American.* You can use one of the indices to the scientific literature such as the *General Science Index,* the *Biological and Agricultural Index,* or the *Index Medicus* if your topic is health related. Index cards are especially useful for recording information from journal articles. Be sure to record information about your reference along with the facts so that you will be able to readily identify the source of important information.

Expand your knowledge by reading some of the references given in your major articles. Also return to the subject headings in the indices of scientific literature to look for more specific information. Use "see" and "see also" references for related materials. If you are not familiar with a particular journal listed in a citation, refer to the journal abbreviations that usually appear in the front of the index.

FEDERAL, STATE, AND LOCAL ENVIRONMENTAL RESOURCES

Many federal government agencies have environmental responsibilities and can provide information about many facets of the environment. These agencies, among them the Environ-mental Protection Agency (EPA) and the U.S. Fish and Wildlife Service, offer a wide range of written materials to the public. Some of these agencies are listed for you in Part III of this book.

Many state and local government agencies, like their federal counterparts, also have environmental responsibilities and also generate a wide variety of environmental information. Some of this material is maintained at public and private university libraries. For local environmental information, try consulting the planning boards that exist in most local government agencies and publicly or privately sponsored environmental organizations. Local sources should appear in your telephone directory or your local library.

PROFESSIONAL, SCIENTIFIC, ENVIRONMENTAL, AND TRADE ORGANIZATIONS

Professional organizations generally specialize in one particular field like engineering or law. These organizations conduct meetings and publish journals. Many have support staff at their headquarters who can help you to locate pertinent information and to identify experts who can provide you with specific information. Scientific organizations consist of scientists in that particular field. They also conduct meetings and publish journals. These materials and personal contacts at local chapters can be important sources of information. Trade organizations are very different; these consist of companies with similar interests. Trade organizations often can provide a wealth of information on environmental issues that are of particular concern to them.

There are many nonprofit environmental organizations worldwide working to preserve the environment. Some of these groups are involved with many different environmental concerns from global climate change to soil loss. Among these are the Environmental Defense Fund (EDF), the Natural Resources Defense Council, Worldwatch Institute, and the Sierra Club. Others, like American Rivers and the Audubon Society, focus on specific issues. Many of these environmental organizations publish newsletters and magazines as well as producing other printed materials. You can contact their national headquarters or a local chapter for information. Some of these groups are included in Part III of this book. The *Nature Directory* lists numerous national as well as grassroots environmental groups:

> *Nature Directory*
> Walker and Company
> 435 Hudson St.
> New York, NY 10014
> (212-727-8300 or 800-289-2553)

RESEARCH CENTERS

Around the country are many nonprofit, university, and government research centers as well as commercial research companies that focus on environmental issues. The research interests of some of these institutions involve global environmental problems and issues while others focus on specific areas: they therefore can be sources for both general and specific environmental information. Part III of this book lists some of the research centers that concentrate on the environment.

THE INTERNET AND WORLD WIDE WEB

The Internet is a global network of computer networks that can be accessed by anyone in the world who has a computer, a modem, and an Internet service provider like AOL, Compuserve, or Netcom. The Internet today offers a massive quantity of environmental information about a wide range of topics. The World Wide Web, in particular, contains numerous sites involving environmental issues and is growing at an astounding rate. The Internet and World Wide Web have become a significant venue for disseminating and sharing environmental information and thus will play a major role in solving environmental problems. You can use the Internet to join environmental discussion groups and newsgroups, to access Web sites, and to view electronic journals and newsletters on the environment. Environmental nonprofit organizations, research centers, university departments, government agencies, and professional and commercial groups host Web sites on the environment. Wherever possible, this book provides Web sites with each idea. Also, Part III lists Web sites of some major environmental organizations and government agencies. For a detailed discussion of how to use the Internet to research environmental topics and an extensive list of Web sites on different environmental issues, consult the following:

> *Environmental Guide to the Internet,*
> 2nd edition, 1996
> Carol Briggs-Erickson and Toni Murphy
> Government Institutes, Inc.
> 4 Research Place, Suite 200
> Rockville, MD 20850

> *Environmental Science and Technology,*
> vol. 30, no. 2, 1996, p. 76A
> Tony Reichhardt

How to Write a Research Paper on the Environment

Whether you are writing a paper or report on the environment or describing your project on the environment, there are some basic procedures that will help you successfully complete your assignment.

Most scientific papers and projects follow the scientific method. Therefore, you must: clearly state your goals; accurately describe your procedure; report and display your results; and then show how your results meet your objectives.

1.   Clearly State Your Goals

Begin your paper by clearly stating your objective(s). It is very helpful if you write this introductory section early on so that you know exactly what you want to do and will be less likely to be sidetracked along the way. In fact, it is a good idea to refer to these goals frequently as you conduct your study so that you remain on course. This introductory section should include background information on your topic along with important references that support the validity of your project idea.

2.   Accurately Describe Your Procedure

Carefully plan your procedure or method before you begin. Write out this section so that you know precisely what procedure(s) you will follow. Make appropriate changes as your project progresses. You may find that procedures need to be altered along the way because you have found more effective techniques or have encountered some unexpected problems. The description that you include with your final report may show the changes that you had to make during your research and experimentation but should make very clear your final procedure. If you conducted a field or laboratory study, accurately describe all techniques and methods used. If your project involved interviews or questionnaires, make sure to include them as well as details on exactly how they were administered.

3. Report Your Results

After you have explained your procedure(s), report your results. Use graphs, tables, and charts wherever possible to summarize your results. Include only what *you* actually found—not the results from another study. Label all tables, graphs, and charts carefully so that the reader can understand the meaning of all given values. It is sometimes helpful to think about how you can show your data before you actually conduct the investigation; this will help you to make clear in your own mind what you are attempting to determine.

4. Discussion of Results

This section is vital to the success of your work. Here you must show how your results relate to your original objectives. What are your conclusions based on your data? Were you able to achieve your objectives? What is the significance of your results? What new problems that might require further investigation are raised by your study? Do your results agree with previous studies or are there discrepancies? Be specific and fully support all of your statements.

SOME GENERAL GUIDELINES FOR WRITING ON THE ENVIRONMENT

1. Be Clear, Accurate, and Direct

When you write about any environmental topic, it is especially important that you state facts and ideas in a direct and clear manner. Environmental topics often overlap with political, economic, and social issues; this means that knowing and understanding the real facts is often extremely important for decision-making. Avoid use of jargon or complicated sentences; your reader must be able to fully understand your work. If you use technical terms, be sure to explain them.

2.   Assume That Your Reader Does Not Know Your Topic

Assume that your reader is an intelligent person with little knowledge of your particular environmental topic. State and explain your ideas clearly and completely. You are better off providing too much information than too little so as to be sure that your reader can understand your work.

3.   Avoid Bias and Generalizations

Appearance of or actual bias is often a major criticism of people who write on the environment. Environmental issues have many facets that affect individuals' lives: protecting forests, for example, has often been viewed by loggers as endangering their jobs; landowners see conservation of wetlands as a way to limit access to their own land; and electric companies oppose stiffer clean air standards because they do not want the initial cost of installing different equipment. Often individuals' needs influence their environmental views. It is therefore critical when you write on the environment that the information you present is reliable and current and that you do not let your personal opinions affect your work.

4.   Include Only Relevant Information

Students often assume incorrectly that a longer report will produce a higher grade. Do not add superfluous information to lengthen your project; include only relevant information to clearly and accurately support your statements. Because environmental science is an interdisciplinary field, it is particularly critical to provide recent, reliable, and complete data to support your statements. Again, you are better off providing too much information in an effort to support your results, but the information must be relevant.

5. Document the Work of Others

Make sure that you document the work of others. Below are some guidelines that will help you determine when you must include references.

- Always use your own words. When you are researching your idea, take notes on each article or section of a book in your own words. If a particular statement is particularly well made, make sure you record it with quotation marks so that you will not inappropriately copy it later as your own.

- Document the experimental results of others as well as their conclusions and explanations of their results. In general, it is not necessary to document facts and explanations that are common knowledge. For example, it is not necessary to document the definition of biodiversity, but it is appropriate to document estimates of the loss of biodiversity in the world in the next 30 years.

- If you plan to include a direct quotation, be careful to accurately reproduce the statement(s).

- *Plagiarism* means representing the words or ideas of another as your own. Thus plagiarism includes having your family or friends write some or all of your paper for you.

- The most common way of referencing someone else's work is to use numerical superscripts following the relevant passages. These references are then listed at the bottom of the page (as footnotes) or on a separate page at the end of the report (as end-notes). Each book reference should include the name(s) of the author(s), the title of the book, the name of the publisher, and the year of publication; each journal reference should include the name(s) of the author(s), the titles of the journal and the article, the volume and number of the issue (if there is one), and the year of publication—and for all published references, the page number from which the information was taken. The particular format of the references will depend on the requirements of your course.

6.   Use Graphs, Charts, and Tables to Display Your Data and Results

Graphs, charts, and tables are very effective ways of showing what you have accomplished. They not only summarize your work but also help your reader understand the significance of your research because they highlight trends, comparisons, and relationships more effectively than do written materials. It is very important, however, to select appropriate types of graphs, charts, or tables. If you are not sure of the appropriate type of illustration to use for your data, consult an instructor or check the literature. Examine articles on similar topics and note how the researchers presented their work. The following reference can also help you decide on how to display your data:

The Visual Display of Quantitative Information, 1983
E. Tufte
Graphics Press
Cheshire, Conn.

If your data consist of precise numbers with several values, tables should be used. These tables, along with graphs and charts, can convey readily recognizable trends and relationships that might be buried if you were to present the values only within the text of your report. Tables should be simple and clear; use brief but precise titles and column headings and appropriate abbreviations and symbols where possible. Make sure the symbols and abbreviations are consistent with your written material and always include units in the column headings where relevant.

TIPS ON CONDUCTING STUDIES ON THE ENVIRONMENT

Below are five general suggestions you should keep in mind as you conduct your study of an environmental issue:

1. Know the Properties of All Reagents and Materials Used

Research the main properties of all materials used, paying close attention to how to use them safely without harming yourself or others or the environment. Look up all reagents in the Merck Index, which is generally kept in the reference section of libraries. Ask your instructors about safe handling of chemicals and supplies and do not use harmful materials in your home. Also, dispose of all materials appropriately; do not, for example, flush harmful organic chemicals down the drain.

2. Learn How to Use All Supplies, Equipment, and Instruments Properly

It is critical that you thoroughly learn how to use all of the supplies, equipment, and instruments before you embark on your project. Improper use will not only lead to unreliable results but may also be dangerous. In addition, scientific instruments tend to be fragile and if misused may require repair and lead to a significant delay in your work. Practice using materials that are necessary for your project before you start. It is always a good idea to talk to instructors or scientists who are familiar with the materials so as to learn from their expertise. Try contacting professors from local colleges; they often are willing to help students involved in environmental projects. You can also contact organizations that are involved with your topic; they generally have a research staff member who may be very willing to assist you. (See Part III of this book, "Resources.")

3.   Understand All Procedures Thoroughly

You should work on understanding all of the procedures that will be included in your project before you even begin. This is as essential for a chemical test as for conducting a survey. There are accepted procedures in every field. Go through several trial runs perfecting your skills. You may find that you will need to alter some parts of the procedure you are using or that you need to make radical changes. Again, discuss your approaches with appropriate individuals.

4.   Decide on How You Will Record and Display Your Data

Determine the best ways to record your data. Construct tables if possible and fill them in as you collect data. Make sure they are accurately and completely labeled so that you will never question the meaning of an entry—never record data on a loose piece of paper that can be easily misplaced. Use one notebook exclusively for your project. Also, think about the best ways to display data in your report. Often if you decide ahead on using a particular kind of chart or graph, you can record your data immediately in a way that will simplify your work later. Examine how other workers in the field have displayed their data and results.

5. Use Only Reliable Sources

Because environmental topics involve so many disciplines and interests, many individuals and organizations with varying degrees of expertise publish material. Use only information coming from credible sources; refer to scientific refereed journals and articles and books written by authors with appropriate credentials. Likewise, when obtaining information from an organization, whether from its publications or from its site on the Internet, make sure that the information has been compiled by knowledgeable staff. Some of the environmental Web sites and pages are created by individuals preparing to enter a field or by individuals interested in, but not particularly well informed about, a particular environmental issue. Numerous errors have appeared in newspapers and magazines on environmental issues, so be particularly selective when researching your environmental topic.

PART II

1001 IDEAS

CHAPTER 1
AIR

Introduction To Air

ABOUT THE ATMOSPHERE

The Earth's atmosphere consists of a tissue-thin layer that protects and nourishes all life. It provides oxygen necessary for respiration and carbon dioxide needed for photosynthesis; it absorbs the sun's harmful ultraviolet (UV) radiation before the rays reach the Earth's surface; and it maintains a fairly constant global temperature, preventing large swings in temperature that would be devastating to many life forms. The atmosphere consists of four major layers: the **troposphere,** the **stratosphere,** the **mesosphere,** and the outermost layer, the **thermosphere.** More than 99% of the total mass of the atmosphere is found within the 30 kilometers (18 miles) closest to the Earth's surface, largely within the troposphere and the lower level of the stratosphere.

The **troposphere** contains most of the air that supports life. It is the first layer above the Earth's surface, extending from sea level to from 10 to 16 kilometers (6 to 10 miles) above the Earth. Air consists primarily of nitrogen gas (78%) and oxygen gas (21%). Other constituents of air include carbon dioxide and argon, which are considered minor components, along with numerous trace gases from both natural and human sources. These trace gases include the wide range of air pollutants from human activities. Water vapor in the air varies as a result of the local climate. The troposphere is also the region of weather, so pilots prefer to fly in the next layer of the atmosphere, the stratosphere. In the stratosphere they avoid the hazards of inclement weather and the turbulence of moving air masses that occurs within the troposphere.

The **stratosphere** protects the Earth from high-energy ultraviolet radiation that is harmful to all plant and animal life. The stratosphere lies directly above the troposphere and is about 30 kilometers (18 miles) in thickness. The stratosphere contains a layer of ozone gas that absorbs destructive UV light from the solar radiation that reaches the Earth's atmosphere. This layer screens about 99% of these high-energy rays.

MAJOR ENVIRONMENTAL ISSUES

Of all the components of the environment—air, water, land, and the living world—the atmosphere has perhaps been most affected by human activities on the most global scale and with potentially the most far-reaching consequences. The unprecedented release of "greenhouse gases," such as carbon dioxide (CO_2), chlorofluorocarbons (CFCs), methane, and nitrous oxide, since the beginning of the industrial age threatens to cause global warming leading to potential major changes in the climate of many regions of the world, as well as raising water levels, and most likely causing more turbulent weather in many nations. There is already some evidence that a warming trend has begun.

CFCs, by virtue of their nonreactive properties, are used primarily as coolants, solvents, and aerosol propellants. Their nonreactvity, however, enables these gases, when released, to rise through the lower atmosphere (troposphere) into the stratosphere to a height of about 15 kilometers (9 miles). Here they react with ozone, a gas that absorbs ultraviolet radiation. The ozone layer in the stratosphere shields the Earth and thus the living world from the harmful high-energy ultraviolet light from the sun. In fact, life cannot exist without the ozone layer's screening out this UV radiation. Although the emissions of CFCs have been significantly reduced through regulated cutbacks in use, the ozone layer continues to thin. CFCs persist in the atmosphere for about a century, so it will be some time before their concentrations return to much lower values. In the 1980s, large ozone holes were detected over Antarctica and then the Arctic; in recent years, studies have shown that the ozone layer is thinning over the populated mid-latitudes of both the Northern and Southern Hemispheres as well. With increased levels of UV radiation reaching the Earth, the risks of skin cancer, cataracts, and suppression of the human immune response system are all increasing. Crop yields and aquatic life are also endangered.

Since the beginning of the industrial age, large quantities of nitrogen and sulfur oxides have been released from the burning of fossil fuels—coal and petroleum products—mainly by electric power plants and internal combustion engines. These gases combine with precipitation to form the "acid rain" that has caused thousands of lakes, in the eastern part of the United States in particular, to become devoid of aquatic life. Acid rain has also contributed to the demise of forests and crops.

The accumulation of nitrogen oxides and hydrocarbons from internal combustion engines causes "photochemical smog" in urban environments. Smog, which forms during daylight hours, requires the sun's energy to produce noxious chemicals like ozone and peroxyacetyl nitrate (PAN). These pollutants are known to cause respiratory problems (they initiate asthma attacks) and even contribute to the risk of lung cancer. Billions of dollars in crop losses in the United States are attributed annually to ozone pollution. This air contaminant has also been shown to play a role in forest decline.

Many studies reported in the 1980s indicated that indoor air pollution often poses a greater threat to human health than does outdoor air pollution. The reality that many people in industrialized nations spend about 90% of their time indoors (home, schools, office buildings, public buildings) and that more energy-efficient buildings reduce air exchange can result in people's exposure to high concentrations of harmful indoor pollutants like radon gas, formaldehyde, benzene, trichloroethylene, asbestos, lead, and the myriad of toxic chemicals present in cigarette smoke.

Global Warming ●

Many scientists believe that the most severe threat to the environment—and therefore to our own welfare—is global warming. In 1992 the United Nations convened its member nations in a special conference on "Environment and Development"; one of the conclusions reached by world scientists at the meeting is that the Earth is warming because of a discernible impact of human activities on the global climate. Vast amounts of carbon dioxide, mainly from the burning of fossil fuels and wood, are spewed into the atmosphere each year. There they trap reflected infrared radiation from the Earth's surface and warm the atmosphere in a process known as the "greenhouse effect." From 1860 to 1990, a little more than a century, atmospheric CO_2 levels increased about 25%, from about 260 parts per million (ppm) to about 350 ppm. And 10% of this increase has occurred in the last 30 years. Other gases like methane, nitrous oxide, and CFCs (chlorofluorocarbon compounds) also add to the "greenhouse effect"; in fact, CFCs are about 20,000 times more effective than CO_2 in absorbing heat.

Estimates vary concerning the rate of warming and the amount of temperature rise. According to the U.N. Intergovernmental Panel on Climate Change, the continued emission of greenhouse gases will cause a global average temperature increase of 1 to 3.5 degrees by the year 2100. The organization also believes that there has already been an impact of human activities on the global climate; the years 1991 through 1995 were warmer than those recorded in any previous five-year period, with 1995 being the warmest year since 1866. The impact of global warming can be potentially devastating for many regions of the world. With rising temperatures, glaciers melt and cause a rise in sea level that will bring about coastal erosion with loss of wetlands, farmland, and living space. Warming is expected to drive the natural hydrological cycle and produce more severe storms. And weather patterns are expected to shift, bringing droughts and floods to regions that have not previously experienced them. In fact, there is considerable evidence today showing rising sea levels from melting glaciers and earlier-starting growing seasons.

IDEAS TO EXPLORE

What models have been used to predict the impact of greenhouse gases on the global temperature? How accurate are they? What major organizations have studied global warming and what conclusions have they reported?

Is there evidence today of global warming? What impact will a rise in temperatures have on global weather patterns? on sea level? on growing seasons? Which regions in the world are expected to be severely affected if greenhouse gas emissions continue at present levels?

What are the potential social, political, and economic problems that may face many nations as a result of the rise in the average global temperatures? For example, will agriculture be affected?

What steps has the international community taken to reduce greenhouse gas emissions? Has the United States agreed to limit greenhouse gases?

PROJECT

The Possible Impact of Global Warming on Your Community ● ○

Global warming is expected to affect the various regions of the world in different ways: some coastal areas are predicted to become inundated (Louisiana, for example, now loses about 64 square miles of land annually from the Mississippi delta into the Gulf of Mexico); wetlands in North America are expected to decrease; growing seasons are even now advancing in the northern hemisphere; and droughts, floods, and severe storms are anticipated in some regions. Research what climate models predict for your state or community as a result of global warming. What impact will these changes in weather patterns have on your community? Will agriculture change or decline? Can the same crops be grown successfully? Is flooding more likely to occur as a result of an increase in the severity and frequency of storms? What are the potential impacts on the indigenous wildlife and plant communities? Will some species disappear while others move in? In past periods of warming, plants and animals moved north with the rising temperature.

PROJECT

Models Used to Predict Effects of Global Warming

✚

Study the computer models used to predict the impact of enhanced carbon dioxide levels and other greenhouse gases on global temperatures and climate. What do these models predict in terms of the future rise in average global temperature? What changes in global weather patterns do they imply? What assumptions are made and are they valid? What are some weaknesses in the various models? How do clouds, ocean currents, and atmospheric reactions add to the uncertainty of predictions based on these models? How do the various predictions compare? For example, do they all point to higher future global temperatures but differ in the magnitude and the rate of increase?

PROJECT

The Impact of Global Warming on the United States ● ○

During the Earth Summit of 1992 in Geneva, 150 nations signed the U.N. Framework Convention on Climate Change. Island nations, Japan in particular, are very concerned over the potential rise in sea level associated with global warming, a process that is already thought to be occurring. Only recently (July 17, 1996), the United States announced that it supports binding targets for reducing greenhouse gas emissions within designated time frames. What are the potential impacts of global warming on the United States? Will agricultural regions decline? How will the cornbelt region, for example, be affected? Will plant communities and wildlife distributions be altered? Will water resources be redistributed? What will happen to cities like New Orleans and states like Florida as ocean levels rise? (More than half of all Americans live within 50 miles of the ocean.) Will the deserts of the Southwest spread as rainfall decreases in the region? Global climate models also indicate considerable increases in the number of hot days in cities like Washington, D.C., and Dallas. How will this impact on city resources? What are some possible economic, political, and social ramifications of global warming? Will some occupations be threatened in some regions while others thrive? Do shifting weather patterns predict greater risks of droughts, flooding, or severe storm events in the various regions of the United States? Use charts and diagrams to outline the various changes in climate and water levels that are predicted to occur in the major regions of the nation. How could the country begin to prepare for these possible changes? The U.N. Intergovernmental Panel on Climate Change (IPCC) has estimated that major greenhouse gases would need to be reduced by 60% immediately in order to stop the rise in greenhouse gas concentrations in the atmosphere. Because of the unlikelihood of such an event, the United States, along with the rest of the world, will most likely see rising average global temperatures as greenhouse gases continue to accumulate at an unprecedented rate.

PROJECT

The Impact of High CO_2 Levels in the Atmosphere and of Global Warming on Local Plant Life ● ○

Plants are basic to all terrestrial ecosystems. Consequently, their response to elevated CO_2 levels, higher average temperatures, and altered moisture patterns will be of utmost importance in ascertaining the impact of climate change on the biosphere. Although controversy still exists regarding the magnitude of the average global temperature rise and climate change, it is certain that CO_2 levels will continue to rise rapidly; in about 50 years, CO_2 concentrations will most likely rise from about 350 ppm at current levels to about 450 ppm. Research has shown that plants respond differently to enhanced CO_2 levels; some plants, like grain, grow more rapidly while others are unaffected. Consequently, those that respond may have a decided advantage over those that do not, resulting in significant changes in the composition of many plant communities. Changes in temperature and rainfall will also result in some plants' competing more effectively. In the past, during periods of increasing temperatures plants and animals tended to migrate to higher latitudes. Study the plant communities in your state or area and research how they may be impacted by rising CO_2 levels as well as changing rainfall patterns and temperatures. Predict what these communities may look like 50 years from now? Use charts and diagrams to show the composition of present and future communities.

PROJECT

The Impact of Global Warming on Indigenous Wildlife ● ○

Study the wildlife that inhabits your state or community or select another area, perhaps a favorite national park. How will elevated temperatures and climatic changes most likely affect this wildlife? Will availability of food and water change? How will changing plant communities affect indigenous wildlife? The foliage of plants enriched by CO_2 has been shown to contain less nitrogen and thus to be less attractive to herbivores. If the population of some herbivores declines, will other species be affected? As grasslands and forests move northward, so will the animals that depend on them. How will the new arrivals interact with those that can stay? Is it likely that some animals will be able to migrate to a suitable habitat?

PROJECT

How to Halt Global Warming ●

In order to eliminate or at least reduce potential future temperature increase and the tremendous dislocation it will cause, atmospheric greenhouse gas concentrations must be decreased. Solutions include conservation, reforestation, energy efficiency, and recycling, as well as shifting to renewable nonpolluting energy. What can each of us do to reduce greenhouse emissions? For one week, study how you add to the atmospheric load of greenhouse gases. You can also determine how much and which gases you contribute. What changes can you make to reduce your impact on global warming?

RESOURCES

Published Materials

Scientific American, vol. 261, no. 3, 1989, p. 70
S. H. Schneider

Scientific American, vol. 273, no. 1, 1995, p. 28
Tim Beardsley

"1995 Is Hottest Year on Record as the Global Trend Resumes," *The New York Times,* Jan. 4, 1996
William K. Stevens

Scientific American, vol. 272, no. 2, p. 13
David Schneider

"For Pacific Islanders, Global Warming Is No Idle Threat," *The New York Times,* March 2, 1997
Nicholas Kristof

Chemical and Engineering News, Aug. 5, 1996, p. 21
Beete Hileman

New Scientist, July 1996, p. 7

Scientific American, vol. 264, 1992, p. 68
F. A. Bazzaz and E. D. Fajer

Bioscience, vol. 39, no. 3, 1989, p. 142
J. P. Cohn

Worldwatch, vol. 2, no. 1, p. 20
J. Jacobson

Web Sites

Climate and Radiation
http://climate.gsfc.nasa.gov/

Earth Observation Center WWW Server Main Menu
http://hdsn.eoc.nasda.go.op/guide/guide/
mainmenu_e.html

Global Change Master Directory (GCMD)
http://gcmd.gsfc.nasa.gov

U.S. Environmental Protection Agency (EPA)
http://www.epa.gov/

Global Change Data and Information System
http://www.gcdis.usgcrp.gov/

Greenhouse Gas Miser Handbook
http://www.ns.doe.ca/co2/greenhouse1.html

United Nations Environment Programme (UNEP)
http://www.unep.ch/

Global Change Research Information Office
http://gcrio.org

SYMBOL KEY

●   Topic of average difficulty

△   Long-term assignment

\#   Project requires special facilities, equipment, or supplies

○   Large public or college library required

*   Safety precautions required

✚   Highly technical; specialized knowledge required

Depletion of the Ozone Layer ●

A thin layer of ozone in the stratosphere, about 20 to 50 kilometers (12 to 30 miles) high, surrounds the Earth, protecting the living world from the sun's harmful ultraviolet (UV) radiation. Ultraviolet radiation (UV-B rays) is known to be both mutagenic and carcinogenic (causes skin cancer), and it suppresses the immune response system as well. In addition, UV radiation can damage several parts of the eye, including the cornea, conjunctiva, retina, and lens, and has a deleterious impact on plants and marine ecosystems. Despite the undeniable dependence of life forms on the ozone layer, human activities have destroyed and continue to destroy it in two main ways: by releasing chlorofluorocarbons (CFCs) into the atmosphere and by flying high-altitude jets that release nitric oxide. Both of these compounds react with ozone molecules in the stratosphere. Presently, we know of two enormous holes in the ozone layer that develop seasonally over the poles (Antarctica and the Arctic). During these periods, abnormally high levels of UV radiation bombard the Earth's surface. Perhaps even more alarming is the gradual thinning of the ozone level at middle and high latitudes (North and South America) where most people live and most of the world's agricultural activities occur. For example, between 1978 and 1990, the ozone layer over North America decreased at a rate of 0.5% per year, so that by 1993, 7.5% of the layer had disappeared. The ozone layer is being destroyed even more rapidly over the Southern Hemisphere. Studies show repeatedly that the UV radiation penetrating the atmosphere is rising with the demise of ozone in the stratosphere. According to NASA's Goddard Space Flight Center, for example, the annual average exposure to harmful UV radiation has increased by 6.8% per decade at 55 degrees north and by 9.9% per decade at 55 degrees south.

IDEAS TO EXPLORE

How do CFCs and nitric oxide destroy the ozone layer? How do these chemicals enter the atmosphere? What are CFCs used for?

Why does the ozone hole appear seasonally over the poles? What causes the ozone layer to deplete over North America?

What happens to the exposure of the living world to harmful UV-B rays as the ozone layer disappears? What impact do these rays have on human health? on plant life? on marine ecosystems? How might agriculture be affected?

Many international agreements have been reached to decrease emissions of CFCs. Have they been successful? It takes CFCs about 15 years to migrate to the stratosphere where they then begin to react with ozone molecules. Also, given that CFCs last about a century and that there are now millions of metric tons of CFCs in the atmosphere, how long will it take for the ozone layer to return to its pre-industrial state?

PROJECT

The Effect of Ultraviolet Radiation on Plants ●
✚

The U.S. Environmental Protection Agency recently stated that UV radiation is harmful to plant growth because it reduces leaf size and thus decreases the surface area needed for energy capture. The preliminary results of two long-term projects researching the impact of increased exposure of plants to UV radiation indicate that wheat and rice, the world's most important food crops, are impaired by UV radiation; growth is stunted because UV radiation decreases photosynthesis. For your project, select several different plants that you can start from seeds and expose them to UV radiation. Monitor plant growth such as leaf size, stem thickness, and height increases, as well as the

quality and quantity of fruit production if you have selected plants like tomato, pepper, or bean. You can also experiment with different exposure periods as well as intensities
of UV light for the different plants (perhaps less exposure
with higher intensities may have a greater effect). For a
long-term project, you could plant the seeds from the irradiated and the unirradiated control plant fruits and measure time for germination as well as the growth of the
plant. In this way you can determine if damage by UV
radiation continues to impact the next generation of
plants. Your project could also involve researching the
impact of UV light on forest trees. Obtain seedlings of different trees and subject them to UV rays. Which ones are
affected? For example, are pine trees more sensitive than
oak trees?

For the above experiments, make sure you have control
groups that use the same soil and are watered and exposed
to sunlight in the same way as the experimental batches.
Keep all variables constant between the different groups of
plants throughout the experiment. Ultraviolet lamps, UV
monitors, and special UV protective glasses are available
through scientific supply companies.

Note: *UV light is harmful to humans, so it is imperative that
you wear protective goggles and carefully follow directions regarding the use of the lamp.*

PROJECT

The Impact of UV Radiation on Seeds ● # ✚

Expose seeds of different plants to UV light before planting. You can use different exposure periods as well as
different intensities. Plant the irradiated seeds and the
controls using the same soil and water them appropriately

(continues)

depending on the type of plant. Is there a difference between the experimental and control batches in the time it takes for germination? Are some species affected while others are not? For a long-term project, monitor the growth of the experimental and control plants to determine if seed irradiation has a long-term effect on plants.

PROJECT

The Effects of UV Radiation on Microorganisms ● # ✚

Studies have shown that some microorganisms are sensitive to increased levels of UV light. Examine measured quantities of pond water under a microscope noting the distribution of the various organisms observed. Subject samples to different exposure periods and intensities of UV radiation and then note any changes in the distribution of organisms as a result of irradiation.

PROJECT

The Interaction of UV Rays and Humic Materials ● # ✚

Humic materials are large organic molecules that come from the decay of plant life. Humic materials are present in all natural water systems—oceans, lakes, rivers, and streams—and give rise to the brownish color of the water. These substances are very important to aquatic ecosystems since they are food sources for plankton and bacteria. During the day, sunlight splits these large molecules into smaller pieces, making them edible to microorganisms. In this project you can research the potential impact of increased levels of UV radiation reaching water surfaces on the integrity of the humic materials. Will they split more rapidly, causing their concentrations to decrease and making less food available to aquatic microorganisms? You can acquire humic acid from a chemical supply company like Aldrich Chemical. Prepare several solutions of humic acid ranging from concentrations of about 0.5 mg/L of organic carbon (typical of concentrations found in seawater) to about 30 mg/L of organic carbon (similar to concentrations in a swamp and accounting for its murky color). Expose your solutions to elevated levels of UV radiation using a UV lamp and appropriate protective eye wear. You can use a spectrophotometer (Spec 20, for example) to look for changes in the humic acids in terms of composition or quantity. You could also use a colorimeter or a nephelometer to monitor the turbidity of the solutions; the lower the concentrations of humics, the lighter the color. Some research has in fact shown that surface water in some oceans is becoming clearer, perhaps caused by increased levels of UV radiation.

RESOURCES

Published Materials

Nature, Oct. 19, 1989
   K. Mopper, University of Miami

Scientific American, vol. 272, no. 1, Jan. 1995, p. 26
   Sasha Nemecek

Geophysics Research Letters, vol. 23, 1996, p. 2117
   J. Herman et al.

Science, vol. 260, 1993, p. 523 J. F. Gleason et al.

Science, vol. 239 (4841 Pt 1), 1988, p. 762
    J. Scotto et al.

Geophysics Research Letters, vol. 22, no. 3, 1995. p. 227
    F. M. Mims III et al.

Science, vol. 262, 1993, p. 1032
    R. A. Kerr and C. T. McElroy

Organizations

NASA's Goddard Space Flight Center

Web Sites

Global Change Mastery Directory
http://gcmd.gsfc.nasa.gov

WWW Server with atmosphere and climate data and information
http://gcmd.gsfc.nasa.gov./pointers/meteo2.html

Atmospheric Radiation Measurement Program (ARM)
http://info.arm.gov

U.S. Environmental Protection Agency (EPA)
http://www.epa.gov
Link: U.S. EPA Stratospheric Ozone
http://www.epa.gov/docs/ozone/
Link: the science of ozone depletion
http://www.epa.gov/ozone/science/science.html
Link: UV health effects
http://www.epa.gov/ozone/uvindex/uvhealth.html

Indoor Air Quality

Because the average American spends 80% to 90% of her or his time indoors, clean indoor air is an important factor in safeguarding public health. Over the last two decades, numerous studies have shown that there are indoor air contaminants that are harmful to health and that should be monitored and reduced. The main indoor air pollutants of concern include radon gas, carbon monoxide, nitrogen dioxide, sulfur dioxide, respirable particles, asbestos, formaldehyde, ozone, lead, and carcinogens and irritants from cigarette smoke.

Radon is a radioactive gas that occurs naturally from the nuclear decay of radium in soil and rock. Radon gas from the earth diffuses through the soil into buildings largely through openings in the foundation. Some building materials and water also contribute to indoor radon levels in some regions. Numerous studies indicate that high levels of radon gas increase rates of lung cancers. When radon itself decays—it has a half life of only 3.8 days—it forms radioactive metals that can become trapped in people's lungs, damaging tissue and leading to lung cancer in some cases. It has been estimated that about 10,000 Americans die annually from radon gas exposure. For that reason, the U.S. Environmental Protection Agency recommends that homeowners measure indoor radon concentrations and take appropriate actions if radon gas levels exceed recommended values.

Formaldehyde is an organic chemical (H_2CO) that is released into the indoor environment from many widely used household items (cosmetics, deodorants, solvents, disinfectants, fumigants, furniture, and carpets) and construction materials (urea-formaldehyde insulation, plywood, and particle board). Reactions to formaldehyde are caused primarily by contact with the skin and the mucous membranes of the eyes, nose, and throat. The health effects appear to depend on the sensitivity of the individual; some people will experience burning eyes and irritation of the upper respiratory passages with concentrations as low as 0.05 ppm (5 parts in 100 million). Concentrations exceeding a few parts per million will cause coughing, constriction of the chest, and wheezing in many individuals, and long-term exposure can cause chronic respiratory disease. Most people, in fact, can detect formaldehyde at 1.0 ppm. The National Institute of Occupational Safety and Health (NIOSH) recommends that no one be exposed to a formaldehyde concentration of 1.0 ppm for any time more than a 30-minute period.

Smoking generates both mainstream and sidestream tobacco smoke. Mainstream smoke results when the smoker inhales and then exhales smoke into the room; sidestream smoke is emitted directly into the indoor environment from the burning tobacco. The passive or involuntary smoker present in the room actually inhales mostly sidestream smoke. Sidestream smoke is more harmful than the smoke inhaled by a smoker because it is produced from a higher combustion temperature and has not been filtered. About 4,000 different substances have been identified in tobacco smoke; over 100 of these are known to be carcinogens. Some of the more toxic pollutants in tobacco smoke include carbon monoxide, benzo-a-pyrene, nitrosamines, nicotine, aldehydes, and acrolein. The latter two chemicals cause eye irritations, nicotine raises blood pressure and heart rate and may be carcinogenic, benzo-a-pyrene and nitrosamines are strong carcinogens, and carbon monoxide binds to hemoglobin, impairing the ability of the blood to pick up oxygen from the lungs. Also, and perhaps most important, carcinogens are carried into the lungs by respirable particles that are now also thought to contribute to asthma. In smoke-filled rooms, the concentration of particulate matter tends to substantially exceed outdoor levels and often is greater than the 24-hour EPA air-quality standard. Recent estimates by the surgeon general indicate that 50,000 cases of lung cancer in the United States are caused by passive or involuntary smoking each year.

Other significant indoor contaminants of concern include combustion products like **carbon monoxide** (CO), **nitrogen** dioxide (NO$_2$), **sulfur dioxide** (SO$_2$), **asbestos** from building materials, and **lead** from paint, plumbing, and outdoor air. Carbon monoxide occurs indoors largely from heating and cooking appliances as well as from tobacco smoke; nitrogen dioxide and sulfur dioxide are both generated by gas stoves; kerosene heaters emit sulfur dioxide in particular. A build-up of carbon monoxide caused by faulty heating or cooking appliances can be deadly; CO is a chemical asphyxiant that has taken the lives of numerous homeowners. Consequently, many people have installed CO monitors in their homes, for use especially during the cold months of the year. Nitrogen and sulfur dioxides are associated with respiratory illness such as bronchitis and asthma. Studies have shown that children exposed to these pollutants experience more respiratory problems than those who are in a cleaner indoor environment. The impact of asbestos on health is well documented: asbestos can cause severe illness like lung and other cancers. Lead also has been studied intensively and is known to have serious health consequences. It has been associated with neurologic, reproductive, and developmental problems as well as with anemia and hypertension. Blood lead levels should be routinely measured in children because high concentrations can cause lower IQ, deficits in speech and language processing, and decreased attention span and behavior problems. (See Chapter 6, "Human Issues.")

IDEAS TO EXPLORE

What are the main contaminants of concern in indoor air? What are their sources and health effects?

To conserve energy, buildings are becoming more airtight? How is the quality of indoor air affected by a low air-exchange rate with outdoor air? How can indoor air quality be improved?

Are there local or state laws that regulate indoor air contaminants in your area? Is smoking permitted in public buildings? What are some of the arguments for and against smoking in public facilities?

Significant numbers of workers have complained of illnesses that they say are coming from the buildings they are working in. Many cases of "sick-building syndrome" have been reported and occur when workers move into new or just-renovated office buildings. Is this illness real or psychological? Are there biological or chemical contaminants present? Is there adequate ventilation?

PROJECT

How Healthy Is the Air Inside Your Home? # ✚

Measure the concentrations of sulfur and nitrogen dioxides, radon, carbon monoxide, and formaldehyde in different areas of your home. Higher levels of NO_2 frequently occur near gas stoves while elevated SO_2 concentrations appear near kerosene heaters. Formaldehyde comes from gases given off by carpets and furniture (outgassing). Radon levels generally are higher where there is less air exchange and where there are holes in the foundation through which radon can enter. For a more inclusive study, you can measure levels at different times of the day and also during different seasons; air quality tends to diminish during the winter months when there is less mixing with outdoor air. You might also measure the levels of these contaminants in the homes of friends and family members.

Many hardware stores sell measuring devices for radon, carbon monoxide, lead, and some for formaldehyde. Sulfur and nitrogen oxides, both indoors and outside, can be determined by spectrophotometric analysis. There are three basic steps to each analysis:

1. Prepare a glass fiber filter impregnated with potassium carbonate. Expose the filter to indoor air for at least a four-week period.

2. Solutions must be prepared for a calibration curve using Beer's Law.

3. Prepare test solutions from your exposed filters, measure their absorbence on a spectrophotometer, and then determine their concentrations from your Beer's Law curve. (For the details of this analysis, see *Ideas, Investigation, and Thought: A General Chemistry Laboratory Manual,* by S. Kennedy et al., Dept. of Chemistry, Hofstra University [Wayne, N.J.: Avery Publishing Group, Inc., 1980].)

PROJECT

Plants That Can Clean Indoor Air ●

Three organic chemicals—formaldehyde, benzene, and trichloroethylene (TCE)—are prevalent in indoor air and can have serious health consequences such as cancer and severe damage to the liver and kidneys. These indoor pollutants come from common household items: formaldehyde emanates from furniture, carpeting, plywood, foam insulation, and paper towels, for example; benzene is released from plastics, rubber, paints, inks, and detergents, to name a few; and TCE gets into indoor air mainly from dry cleaning but also is present in adhesives, varnishes, and paints. According to a 1989 report by the National Aeronautics and Space Administration (NASA), we can reduce our exposure to these toxic chemicals by keeping certain plants indoors. In fact, their report indicates that certain plants could significantly reduce the

(continues)

concentrations of these air contaminants, in some cases by as much as 90%. For example, common household plants like philodendron, spider plant, aloe vera, snake plant, bamboo, date palm, azalea, and poinsettia can bring down formaldehyde levels, while the dragon plant and dracaena warnecki (variegated) lower benzene and TCE concentrations.

For your project you could monitor the concentrations of formaldehyde in different areas of your home. You might begin by looking at a room where there is new furniture or carpeting; outgassing of formaldehyde and other organic compounds tends to be highest when the items are new. Bring in plants that are known to reduce formaldehyde levels. Make sure the plants have the necessary lighting required for their growth. Two or three plants should be in every 100 square feet of floor space for the average eight-foot-ceiling room. Did formaldehyde levels drop? How long did it take? Try other plants to see if they have an impact on concentrations.

RESOURCES

Organizations

National Aeronautics and Space Administration
  B. C. Wolverton

U.S. Environmental Protection Agency

Consumer Product Safety Commission (CPSC)

Occupational Safety and Health Administration (OSHA)

National Institute of Occupational Safety and Health (NIOSH)

Centers for Disease Control and Prevention (CDC)

U.S. Public Health Service

U.S. Department of Health and Urban Development

Published Materials

Redbook, Sept. 1992, p. 69
  Joel Rapp

Indoor Air Pollution, 1983
  Richard A. Wadden and Peter A. Scheff
  John Wiley and Sons, New York

Radon and Its Decay Products in Indoor Air, 1988
  Edited by W. W. Nazaroff and A. V. Nero, Jr.
  John Wiley and Sons, New York

New England Journal of Medicine, vol. 319, Dec. 1988,
  p. 1452 J. Fielding et al.

Human Ecology, Winter 1992, p. 2
  Alan Hedge

Many articles on indoor air contaminants have appeared
in *Science, Environmental Science and Technology*, the *New
England Journal of Medicine*, and *Chemical and Engineering
News*. Also, an independent trade newspaper, *Indoor
Environment Review*, published monthly, is a valuable resource
(800-394-0115).

Web Sites

Environmental Protection Agency's Indoor Air Quality Home Page
http://www.epa.gov/iaq/
Link: Indoor Air Quality Info Clearinghouse
http://www.epa.gov/iaq/iaqinfo.html
Link: More Information on Air
Qualitywww.epa.gov/iaq/moreinfo.html

PROJECT NOTES

Sick-Building Syndrome (SBS) •

Most people in the United States spend 80% to 90% of their time indoors, with a significant portion of that time at work. In the last decade in particular, jobs have shifted from manufacturing to service-oriented fields, placing many workers inside large office buildings, many of which are not adequately or appropriately ventilated. Many outbreaks of illnesses have been reported that are associated with some buildings. "Sick-building syndrome" (SBS) refers to a wide range of symptoms including dizziness, nausea, lethargy, headache, and skin and mucous membrane irritations that improve when the affected worker leaves the building. It has been estimated that employers sustain losses of $65 billion annually as a result of SBS; the illness causes high worker absenteeism as well as lower productivity.

SBS frequently occurs in air-conditioned buildings. It frequently occurs when workers are moved to new or renovated buildings in which outgassing of volatile organic compounds (VOCs) occurs. Consequently, indoor air contaminants are suspected of causing the reported symptoms. Some Finnish studies show worsening of symptoms with increases of temperature, perhaps as a result of higher VOC levels. Increasing ventilation rates in sick buildings has also been shown to aggravate the illness; increased ventilation is thought to bring in more contaminated air, especially if fungi and dust mites are present. These biological contaminants are notorious for causing an array of allergic responses. In many cases no particular indoor contaminant(s) can be identified, thus indicating that other factors may play an important role in SBS. These include occupational factors like hours on a computer, lighting, noise levels, job stress, and job satisfaction. It is becoming clear that psychological issues may affect a worker's susceptibility to SBS.

IDEAS TO EXPLORE

How prevalent is "sick-building syndrome" in the United States? How about in other countries?

What major studies have been reported that investigated the cause(s) of SBS? What factors play a significant role in the development of the illness? Look at some specific cases of a sick building and the quality of its indoor environment? Are complaints more prevalent in areas where there are copiers, new carpets and furniture, adhesives, and other supplies that release organic compounds? What are the noise levels, and how adequate is the lighting? How important are psychological and occupational factors?

What remedies have been applied, and were they effective?

PROJECT

Are There Sick Buildings in Your Community? ●

Are there buildings in your area in which occupants feel sick when inside? What are their complaints? Are they exposed to office equipment and supplies that release noxious organic compounds? Is there new carpeting or furniture? How high is the noise level and how adequate is the lighting? Do workers who spend more time in front of a computer complain more frequently? Inquire about job stress? How often do workers miss work because of their symptoms? Identify some of the issues that appear to be related to the complaints or symptoms.

PROJECT

School Indoor Environment and Student Performance ●

Study the indoor environment of your school and its impact on how much you learn. Do you or any of your classmates show symptoms of SBS: headache; lethargy; dizziness; eye, ear, or nose irritation; itchy skin; chest tightness; or difficulty in breathing? Do these symptoms disappear when you leave school? Are they related to air-conditioning or to high indoor temperatures or humidity? The southeastern United States tends to have high concentrations of fungi in indoor air. Is there adequate air exchange with the outdoors? Do you learn more effectively when windows and doors are opened? What materials release noxious gases? Is the lighting adequate? Crowded rooms mean more carbon dioxide in the air, heat, humidity, and many odors and contaminants from dry-cleaned clothes, perfumes, and deodorants. Do you develop symptoms especially under crowded conditions? You can design a survey for your classmates to complete investigating the health of your school's indoor environment and its impact on learning.

RESOURCES

Published Materials

Human Ecology, Winter 1992, p. 2
   Alan Hedge

Indoor Air Review, Apr. 1996, p. 3

Organizations

National Institute for Occupational Safety and Health (NIOSH)

Environmental Defense Fund

Indoor Air Quality Information Clearinghouse (sponsored by the U.S. EPA)
P.O. Box 37133
Washington, DC 20013-7133
(202-484-1307 or 800-438-4318)

Public Relations Office American Society of Heating, Refrigerating and Air-Conditioning Engineers (ASHRAE)
1791 Tullie Circle NE Atlanta, GA 30329

Also, you can contact your state or local health department.

Web Sites

Sick-Building Syndrome
http://www.epa.gov/iaq/pubs/sbs.txt

SYMBOL KEY

● Topic of average difficulty

△ Long-term assignment

\# Project requires special facilities, equipment, or supplies

○ Large public or college library required

✳ Safety precautions required

✚ Highly technical; specialized knowledge required

Air Pollution and Acid Rain ●

Two gaseous air pollutants, sulfur and nitrogen oxides, are posing a universal threat to the environment. These gases combine with water to produce acid deposition—commonly called acid rain—which turns lakes acidic thereby killing aquatic life, damages crops, decimates forests, decreases soil fertility, and destroys statues and buildings, many of which are of historical significance. Although sulfur and nitrogen oxides do emanate from some natural sources like volcanoes, most of these acid precursors in eastern North America and in many other regions of the world result from human activities. In eastern North America, for example, more than 90% of sulfur oxides and 95% of nitrogen oxides are of human origin. Most of the sulfur oxides are released by electric power plants, the majority of which burn coal. The main anthropogenic sources of nitrogen oxides are electric power plants and motor vehicles. These gases enter the atmosphere and often travel long distances until they can combine with precipitation to form acid rain, snow, or fog; sometimes they settle out of air like dust particles (dry deposition) and then react with water on surfaces to form acids.

IDEAS TO EXPLORE

What causes acid deposition? Distinguish between dry and wet deposition?

Where is it happening and why? What are the environmental impacts of acid deposition?

Which human activities are most responsible? How do the tall stacks of electric power plants add to the problem? Which areas of the United States have been most affected and why?

What are some solutions? What is being done in other nations facing the serious problems of acid rain? What technology is now available to reduce these acidic precursors at their source? How will improvements in energy efficiency help solve the problem?

PROJECT

The Impact of Acid Deposition on Natural Water Systems ● ○

Acid deposition is occurring worldwide. Developing nations with weak environmental regulations—or no environmental regulations at all—are creating high levels of acid precursors in their urban areas of heavy traffic congestion and concentrated industry. Lakes, rivers, aquatic life, and wildlife are succumbing to acid deposition in every portion of the world. Thousands of lakes and streams in the eastern United States have been acidified and are now devoid of life, in Sweden about 20,000 lakes are dying or already dead, and nine of Nova Scotia's rivers have lost their salmon populations. Other wildlife are suffering as well. Songbirds living near these acidified lakes are producing eggs with soft shells, and many amphibian species are disappearing because of acid deposition. The decline of species in any ecosystem often impacts on the ability of the whole system to survive.

Select a region of the world, the United States, or your home state and research the lakes, rivers, and streams that have been acidified and the impact of acidification on the indigenous aquatic and nearby wildlife. (The National Wildlife Federation has a list of U.S. lakes that have become acidified.) Use charts, graphs, and tables to show the decline by plotting dropping pH values. Determine the major sources of the acid precursors, nitrogen and sulfur oxides, for your selected region. How do wind currents play a role in transporting pollutants to your area?

PROJECT

The Impact of Acid Deposition on Crops, Forests, and Soil ● ○

Damage to crops and trees by acid precipitation is thought to cost billions of dollars. Acids appear to damage leaves of many plants, affect normal bud development, and alter soil chemistry causing calcium and other nutrients to leach out. A multitude of studies have shown that acid precipitation is decimating many forest systems throughout the world. Birch, pine, spruce, and oak trees have been destroyed in many forest regions; half of the red spruce trees in the Green Mountains of Vermont have died from acid precipitation.

Select an area—your own state, or a park like the Adirondack State Park of New York, or perhaps a specific forest like the Black Forest of Germany—where trees are dying. Research the impact of acid rain over the last decade or so on the health of the forest ecosystem. Which trees are in trouble? What mechanisms may be playing a role in their decline? Where is the pollution coming from? Are other pollutants also having a deleterious effect on the health of the forest? Show correlations between years of exposure to acid rain and forest decline.

PROJECT

Measure Concentrations of Sulfur and Nitrogen Oxides in Your Community # ✳ ✚

You can measure concentrations of sulfur and nitrogen oxides in outdoor air by spectrophotometric analysis. There are three basic procedures for each analysis.

(continues)

1. Prepare a glass-fiber filter impregnated with potassium carbonate. Keep this filter in a designated location for a four-week period.

2. Prepare solutions for a calibration curve using Beer's Law.

3. Prepare test solutions from the filters, measure absorbence of these solutions using a spectrophotometer, and then use your Beer's Law curve to find their concentrations. (For the details of this analysis, consult *Ideas, Investigation, and Thought: A General Chemistry Laboratory Manual,* by S. Kennedy et al., Dept. of Chemistry, Hofstra University [Wayne, N.J.: Avery Publishing Group, Inc., 1980].)

Measure concentrations in several locations, during the different seasons, and at various hours of the day. Are concentrations of these acid precursors greater at night or during the day? Do they differ according to season? Are nitrogen oxide levels greater in areas of high traffic volume?

RESOURCES

Published Materials

Environment, vol. 28, no. 4, 1986, p. 6
  A. H. Johnson

Last Stand of the Red Spruce, 1987 R. A. Mello
  Natural Resources Defense Council and Island Press

World Resources, A Report by the World Resources Institute, 1992–1993, 1994
  Oxford University Press, New York

Scientific American, Sept. 1990, p. 60
P. Crutzen and T. Graedel

The Chemistry of the Atmosphere: Its Impact on Global Change, 1993
Edited by John W. Birks, Jack G. Calvert, and Robert E. Sievers
American Chemical Society, Washington, D.C.

Numerous articles in Environmental Science and Technology, Science, and *Chemical and Engineering News*

Organizations

U.S. EPA, Office of Air Quality Planning and StandardsResearch Triangle Park, N.C.

Web Sites

EPA Office of Air and Radiation
http://www.epa.gov/oar/
Link: http://www.epa.gov/acidrain/ardhome.html
Link: http://www.epa.gov/acidrain/envrben.html

IGC Resources (many links)
http://www.econet.apc.org/acidrain/

Acid Rain
http://www.ns.ec.gc.ca/aeb/ssd/acid/acidfaq.html

PROJECT NOTES

Urban Smog I ●

Today in America nearly four in ten people suffer from respiratory problems caused by urban smog. This means that 100 million residents experience health problems because they breathe polluted air. According to the American Lung Association, 64,000 Americans die prematurely from cardiovascular and respiratory problems caused or aggravated by fine particulate pollution. Children in particular are at risk because they generally spend much more time outdoors and are much more susceptible than adults to respiratory problems induced by air pollution.

Air pollution comes from three main sources: transportation, combustion of fossil fuels (mostly in electric power plants and factories), and from a variety of industrial processes. When hydrocarbons and nitrogen oxides (mostly from automobiles and power plants) mix in polluted air in the presence of sunlight, other contaminants (called secondary pollutants) are formed. These give rise to urban or photochemical smog. Ozone, formaldehyde, and peroxyacylnitrate (PAN) are among the most harmful secondary pollutants. Ozone is one of the most prevalent pollutants in smog, and one of the most damaging. Ozone irritates the respiratory system, destroys trees and crops, and erodes rubber. In fact, a joint study by the U.S. Environmental Protection Agency and the U.S. Department of Agriculture indicates that the yearly crop harvest in the United States is reduced by at least 5%, and possibly by as much as 10%, because of air pollution, resulting in a loss of $3.5 to $7 billion annually.

Air pollution has become a very costly environmental problem both in terms of human lives and in terms of billions of dollars lost in health-care expenditures and crop damage. To reduce the harmful impacts of air pollution on society, the EPA is currently attempting to update the National Ambient Air Quality Standards by placing tougher regulations on ground-level ozone limits and by, for the first time, regulating small particles. Although numerous studies have indicated that illness and death rates are associated with airborne fine particulate matter (2.5 to 1.0 microns in size), these are presently unregulated and are being emitted into the air with impunity. Many industrialized nations have markedly reduced air pollution by decreasing their use of fossil fuels through proper insulation of homes and buildings, for example, and by using energy more efficiently.

IDEAS TO EXPLORE

How does urban or photochemical smog form? What are the sources of the primary pollutants that cause smog? Which secondary pollutants are most dangerous?

Fine particles in smog have been shown to cost thousands of Americans their lives each year. How do these small particles form and why are they so dangerous?

How can air pollution be reduced? What should be done by government agencies, industry, and the public?

PROJECT

How Healthy is the Air in Your Community? ● ○

Contact the local pollution control authorities in your community or consult local newspapers for the levels of air pollutants. You might chart ground-level ozone concentrations over several months or track these over longer periods by consulting previous reports. During which seasons are levels the greatest? During which time of the day? How do weather conditions impact air pollution? How does your community compare to others? During how many days does the air not meet health standards? Even if you do not live in a city, keep in mind that often suburbs and rural regions have even higher levels of air pollution than urban centers because polluted air often drifts out of the city into outlying areas.

RESOURCES

Published Materials

Breath Taking, 1996
A report by the Natural Resources Defense Council (NRDC)

"More Than 50,000 Die Each Year From Air Pollution,
Environmental Group Reports,"
The New York Times, May 9, 1996
Phillip J. Hilts

New England Journal of Medicine, vol. 329, 1993, p. 1753
D. W. Dockery et al.

Organizations

U.S. Environmental Protection Agency (EPA)

American Lung Association

Web Sites

Office of Air and Radiation: Office of Mobile Sources
http://www.epa.gov/OMSWWW/
Links: http://www.epa.gov/OMSWWW/o5-autos.html

Air Radiation and Toxics Division (ARTD)
http://www.epa.gov/reg3artd
Links: Clean Cities
http://www.epa.gov/reg3artd/partner/clncit.html
Links: Protecting Our Air Quality
http://www.epa.gov/reg3artd/airqual/hmpg.html

CHAPTER 2
LAND

Introduction to Land

ABOUT THE LAND

Only about 29% of the Earth's surface is covered by land; the rest of the surface is mostly salt water, primarily in the oceans. Almost all of terrestrial life depends on the thin layer of soil that coats much of the Earth's land. Civilization, in fact, would collapse without fertile soil. Although we all depend on the health of soil, we tend to think of it as dirt needing only doses of fertilizer to keep it producing crops. Soil is, however, a vast and very complex ecosystem teeming with life; one gram of fertile agricultural soil contains about 30,000 one-celled animals, 50,000 algae, 400,000 fungi, and 2.5 billion bacteria. Earthworms, roundworms, tiny insects, and mites also live in soil. The fertility of soil depends on these living inhabitants as well as on many complex physical and chemical processes. Inorganic nutrients in the soil come from the breakdown (weathering) of rocks through many different chemical and physical processes, while organic components (humus) come from the decomposition of animal and plant life. Soil depth ranges from only a few centimeters in mountainous, rocky, or arctic regions to sometimes a meter or two in the fertile forests or prairies. Soil formation is a very complex process requiring many physical, chemical, and biological processes operating over hundreds or even thousands of years.

MAJOR ENVIRONMENTAL ISSUES

In the natural world, soil formation and soil erosion—the removal of soil, usually by wind or flowing water—occur at about the same rate, making soil a renewable resource. However, our actions, in particular our ever-increasing need for more cropland to feed our exploding numbers, our use of poor agricultural practices, and our increased use of water resources for irrigation have led to the rapid loss and degradation of soil in many regions of the world. Topsoil today is eroding faster than it forms on approximately one-third of the world's cropland. Topsoil erosion is associated with 85% of the world's land deterioration. It is estimated that the planet is losing 7% of its topsoil from cropland or potential cropland each decade.

Approximately 10,000 years ago, when humankind began agricultural activities, human settlements expanded. A cycle of increasing population followed by the need for more food and water, mostly to irrigate crops, commenced at that time and continues today. The Earth's surface has changed significantly over the last three centuries as a result. On all of the continents, humans have cleared forests, drained wetlands, and irrigated grasslands, mostly for the purpose of creating farmland. It has been estimated that from 1700 to 1980 cropland increased by about 460% while forest and woodland declined nearly 20%. In recent decades the rate of cropland expansion has risen dramatically so that more land has been allocated for crops in the 30-year span from 1950 to 1980 than in the 150-year span from 1700 to 1850.

Human activities have markedly altered the Earth's surface. By the early 1990s about 40% of the world's land surface had been changed to farmland and pastures in order to provide food for our exploding population. This enormous land transformation has occurred primarily at the expense of forests and grasslands. Tropical forests and global soil resources in particular have been affected during the last half century with the doubling of the world's population. It has been estimated by the United Nations Environment Programme (UNEP) that about 11% of the Earth's vegetated soils—an area the size of China and India combined—have become severely degraded so that they can no longer sustain plant life. Global deforestation has reduced the world's forests by one-quarter. The rate of tropical forest destruction increased rapidly in the 1980s so that today only about one-half of these vital resources remains. Each year about 17 million hectares (about the size of Florida) of tropical forests are clear-cut, and land the size of Great Britain is turned to desert, no longer capable of supporting vegetation. Devastation of the land is particularly acute in the poor countries of Africa and Asia and in Central and South America. It is important to note that today much of the land cleared in the non-industrial countries is used to grow grain for livestock, particularly cattle. This is a marked change from the recent past when most of the grain produced was consumed by people. The meat from the cattle is largely sold to the industrial nations for consumption, often leaving people of these cash-poor nations without sufficient food and with the environmental devastation associated with cattle grazing and poor farming practices as well. By providing meat products to a very small but wealthy percentage of people, poor nations deprive their own people of the ability to feed themselves.

Pollution has also impacted on the quality of the land. Release of toxic wastes has left many areas in the world unoccupied. The release of enormous amounts of radioactive materials from the Chernobyl nuclear power plant in Ukraine, for example, caused the evacuation of hundreds of thousands of people from the land around Chernobyl with no hope of ever returning. The world's most severe industrial accident occurred at Bhopal, India, at a Union Carbide pesticide plant. About 40 tons of a toxic gas exploded into the atmosphere, killing thousands of Indians and seriously maiming over 100,000 people. In the United States, chemical spills and leakage from landfills have also claimed many areas. For example, in the 1970s residents of Love Canal, a large housing development in New York State, were forced to evacuate because of harmful chemicals that were leaking from a hazardous waste site. Air, water, and soil were contaminated with toxic and carcinogenic chemicals, exposing hundreds of people to increased risk of cancers, birth defects, and respiratory, kidney, and nerve disorders.

Global Deforestation •

Before the advent of agriculture about 10,000 years ago, forests covered approximately one-third of the Earth's surface; today they occupy only one-quarter of the world's land area. That means that our activities have reduced forests by 25%; of what remains, only about 12% are still in their natural, undisturbed state. The trend continues with forests disappearing and being fragmented in many parts of the world as more land is cleared for farms and timber to produce food and housing for the world's exploding population.

Aside from providing timber, fuel, and many commercial products, forests play a major role in stabilizing the world's climate, maintaining the atmosphere, purifying water and air, preventing floods, and sustaining a diverse array of animal and plant life. In fact, tropical rain forests alone are thought to harbor about one-half of the world's species. According to the Food and Agriculture Organization of the United Nations (FAO), acute deforestation is occurring in these vital ecosystems. If the current rate of tropical forest destruction continues, they will just about disappear in 30 to 50 years. Tropical forests are particularly vital to the health of the global environment, and the consequences of their loss to the world's air, water, biodiversity, and climate can be catastrophic.

Old-growth forests in the Pacific Northwest and Canada are also rapidly disappearing. Old-growth forests are home to many different species that require their large stands of dead trees and fallen logs.

IDEAS TO EXPLORE

What is the ecological value of forests? How do they protect our water resources? How do they prevent flooding? How do they purify air and stabilize the world's climate?

Determine the many commercial products that are extracted from forests.

What is the difference between natural forests and timber farms? What are the disadvantages and advantages to using monoculture tree plantations? Where are forests disappearing most rapidly?

Where are the world's tropical forests located and where are they most rapidly disappearing? Why are tropical forests so valuable? What are the causes and possible consequences of this massive destruction of such a vital ecosystem?

What has happened to the temperate and boreal forests of the world since the industrial revolution? Describe more recent trends. What has been the impact of the overall loss of forests in Europe and the Northern Hemisphere? Why are old-growth forests so valuable?

How can forests be used as a sustainable resource? Give some examples of how forests are being economically beneficial without being destroyed. What are some long-term economic disadvantages to clear-cutting forests?

RESOURCES

Published Materials

World Resources, a Guide to the Global Environment,
    1995–1996
    Oxford University Press, New York

"Reforesting the Earth," in *State of the World,* 1988
    S. Postel and L. Heise
    W. W. Norton and Co., New York

Organizations

United Nations Food and Agriculture Organization

United Nations Environment Programme

Deforestation in the U.S. ●

Today America has lost about one-third of its forests, and about 90% of America's original forests have been cut. As settlers worked their way across the land, clearing wilderness for farms and homes, they first depleted the East of forest cover and then continued on to the West. Although forests have grown since the turn of the twentieth century, they are mostly second- or third-growth forests. Many of these are timber farms hosting only one species of commercially valuable tree. These monocultures lack the biodiversity of naturally occurring forests and therefore are more prone to pest infestation, disease, and the deleterious effects of air pollution. Most of the remaining ancient or old-growth forests are located on public lands in the Pacific Northwest—Washington, Oregon, and northern California—where they are rapidly being clear-cut by the timber industry for short-term economic gain.

You can use charts and graphs to show changes in the quantity and quality of forests in the United States.

IDEAS TO EXPLORE

How did America look before the arrival of the Europeans? When and why did many forests disappear? What are the current trends? There is some evidence that forests are actually increasing in temperate regions. Is this true for the United States?

What is the difference between undisturbed ancient forests and second- and third-growth forests?

What has been the impact of deforestation on soil erosion, water quality, and flooding? What regions of the United States have been most affected?

What has happened to our old-growth forests? Why are they, like the tropical rain forests, being destroyed so rapidly? Why are they particularly important?

How can these important resources be preserved? Suggest some solutions.

PROJECT

Select and Study a Forest in Your Region, State, or Community ● ○

Select a region of the country (Southwest, Northeast, etc.), state, or smaller area and track deforestation. How did the land look to Native Americans 300 years ago? What species of trees, smaller plants, and animals were indigenous to these ancient forests? Trace the changes in the forested areas. What were the causes for the changes in the forested areas? Were forests removed for logging or cleared for farmland or housing? What was the impact on water resources? soil? biodiversity? What species of animals and plants were affected? Were the ancient forests replanted? How do the second- or third-growth forests differ from the old-growth forests? How can forests be used sustainably? Propose some solutions.

PROJECT

Reducing Your Use of Forest Products and Saving Trees ●

Pressure on forests can be relieved by increasing efficiency as well as by decreasing demand by consumers. According to the U.S. Forest Service, homes could be constructed safely using 10% less lumber, and improved machinery could decrease the amount of wood needed for plywood by one-third. Consumers can also dramatically reduce their use of wood products. In the United States, each of us consumes over 600 lb. of wood annually in lumber and paper as compared to about 110 lb. consumed annually by each European and 15 lb. by people in Third World countries.

To ascertain how you can reduce your reliance on forest products, record all forest products used by your family or yourself in one month. Remember that there are about 5,000 different products that come from felling trees, so carefully research the source of the products and goods you use. Include only those products that are obtained by cutting trees—not products like maple syrup or nuts and berries that actually encourage good forestry practices. Products of interest here include timber, pulp and paper products, fuel, wood alcohol, charcoal, adhesives, and many pharmaceuticals. Determine how you can reduce your use of these products to reduce logging, especially in the tropical rain forests and old-growth forests. For example, what can you substitute, reuse, recycle, or use less of? Can you use cloth shopping bags, use cloth or sponges in place of paper towels, use e-mail, reuse lumber, use recycled paper, buy king-size products reducing wrapping waste, etc.? Then estimate by how much you have decreased your use of forest products and estimate the number of trees you can save. Reducing use of paper products will also markedly reduce solid wastes that are overburdening landfills.

RESOURCES

Published Materials

Mountain in the Clouds, 1982
    B. Brown
    Simon and Schuster, New York

The Forest and the Trees: A Guide to Excellent Forestry, 1992
    G. Robinson
    Island Press, Washington, D.C.

Rainforest in Your Kitchen, 1992
  M. Teitel
  Island Press, Washington, D.C.

Organizations

U. S. Forest Service

Wilderness Society

Web Sites

Forest System Ecosystem Management
http://www.fs.fed.us/land/welcome.htm

World Resources Institute
http://www.wri.org
Links: Forest Resources
http://www.wri.org/wri/biodiv/foresthm.html

Gaia Forest Archives
http://gaia1.ies.wisc.edu/research/pngfores/gaia.html

Environmental Defense Fund
http://www.sun-angel.com/edf/edf.html

SYMBOL KEY

● Topic of average difficulty

△ Long-term assignment

\# Project requires special facilities, equipment, or supplies

○ Large public or college library required

✳ Safety precautions required

✚ Highly technical; specialized knowledge required

The Old-Growth Forests of North America ● ○

Old-growth forests are forests that have not been significantly disturbed for hundreds of years. In old-growth forests, conifers like Douglas fir, giant sequoia, and the western hemlock are known to live for 500 years, and some can survive for 3,000 years. In the United States only about 10% of these original forests remain. Although some loblolly pine forests exist in the Southeast, most old-growth forests in the United States are located in the Pacific Northwest.

Old-growth forests, like tropical forests, are valuable because of their ecological as well as their commercial importance. Because old-growth forests contain huge numbers of standing dead trees as well as logs, they provide a unique habitat for many diverse species. At this point we have barely begun to study the numerous species that inhabit these ancient forests. Some of these may harbor anticancer medicines, new plant crops, or genes to make crop plants more resistant to pests, reducing the need for pesticide applications. In addition to their significant contribution to global biodiversity, old-growth forests soak up tremendous quantities of water from heavy rainfall events and snow melts, recharging groundwater and delivering purified water to lakes, rivers, and streams. Their contribution to the health of water resources in the Northwest is vital for the survival of the fishing industry. In fact, heavy clear-cutting of old-growth forests has contributed to a significant decline in salmon stocks as rainfall events drag massive quantities of sediment from bare regions into streams and rivers. The sediment suffocates spawning beds and interferes with young salmon. As a result of this excessive logging, whole salmon runs have disappeared, impacting on the economy of the region where the commercial salmon-fishing industry employs over 60,000 people and attracts about $1 billion annually.

Aside from the ecological services they provide, old-growth forests are of tremendous commercial value, particularly to the timber industry, which is destroying old-growth forests at a rate exceeding the demise of tropical forests. If the present rate of clear-cutting continues, most will have disappeared in 20 to 30 years. Clearly, the timber industry is exchanging short-term windfall profits for the destruction of a unique forest system that in only a short time will no longer be capable of supporting the industry, along with its 100,000 employees, nor of fulfilling its vital ecological services to the region. The major difficulty in preserving these forests lies in the needs of the thousands of local residents who rely on the timber industry for their livelihoods and of those who benefit from the millions of dollars that the timber industry contributes to the region's economy. If ancient forests and the wild plants and animals are to be saved, solutions must involve new jobs for timber workers; the new jobs must relate to activities that support the sustainable commercial use of the forests such as tourism, selective cutting, restoration of degraded areas, and job training. In addition, industries should pay for the environmental damage their activities incur, including flooding from clear-cuts, demise of fishing from sediment runoff, and water pollution.

IDEAS TO EXPLORE

Where were the ancient forests located in the United States before the arrival of Europeans on the continent? Where are they located today?

Why are old-growth forests important to all of us? Why in particular to residents of the Pacific Northwest?

The Endangered Species Act has been used to spare some of the forests in order to save the spotted owl from extinction. (The northern spotted owl has been listed as a threatened species.) Loggers and others feel that in enforcing this act the government is elevating the importance of the survival of this owl above their own survival and well-being. Is this really the issue? What are some solutions? How can the forests be used sustainably to benefit the local residents and the rest of the population?

PROJECT

The Controversy over the Old-Growth Forests of the Pacific Northwest and the Spotted Owl ● ○

With the placing of the northern spotted owl on the Threatened Species list, one of the most controversial and publicized environmental problems involving the old-growth forests arose. According to the Endangered Species Act, the habitat of threatened species—that is, species whose population is declining so rapidly that it has become endangered—must be saved. As a result, some of the old-growth forests in the Pacific Northwest were removed from the ranks of forestland that the timber industry could exploit. The industry warned that 50,000 jobs would be lost as a result of protecting the owl and raised the philosophical issue of elevating the protection of wildlife over the needs of human beings. Workers in the many towns of the Northwest that depend on the timber industry expressed anger and resentment over their potential job loss caused by what they perceived as placing the needs of an owl over their own welfare.

Proponents of the protection of the spotted owl, on the other hand, emphasized the need to protect an ecosystem that is being decimated at an exploding rate. Destruction of the ecosystem will, in a relatively short period of time, lead to the loss of many more jobs as a result of the disappearance of the old-growth forests. They point out that the timber industry has poorly managed America's native forests. In only 150 years old-growth forests have been reduced to 5% to 10% of their initial land cover and the

(continues)

remaining ancient forests are severely fragmented from clear-cuts. At this time, old-growth forests are being harvested at twice the rate at which Brazil is harvesting its tropical rain forests. The argument that preserving some forestland will eliminate jobs is misleading; automation and exporting of raw timber are primarily behind the recent loss of jobs. In fact, in several logging regions of the Northwest, increased production was actually accompanied by job loss—not gain—as a result of increased automation and exporting raw timber to nations like Japan that then use their own timber mills. The number of jobs offered to workers in the logging regions will continue to decline as automation replaces people and profits from exports continue. Many believe that trees cut from federal forests should be used for domestic consumption and that these trees should be processed here so that jobs are not also exported abroad.

Collect various articles from newspapers from different regions of the nation, including local as well as national papers like the *New York Times* and the *Chicago Tribune*. How do they present the problem? Do the local papers from Oregon, for example, differ markedly from the more national newspapers? What are the real issues here? Is it really a case of jobs versus the spotted owl? Write your own article about this important environmental issue including suggestions of how it can be solved. According to E. Goodstein, only about 0.1% of all layoffs in the United States are due to environmental regulation; jobs are lost to other countries because of their lower labor costs, not because of less environmental regulation.

RESOURCES

Published Materials

Wilderness, vol. 52, no. 182, 1988, p. 56
T. H. Watkins

The Forest and the Trees: A Guide to Excellent Forestry, 1988
G. Robinson
Island Press, Washington, D.C.

Environmental Management, vol. 20, no. 3, 1996, p. 313
E. Goodstein

Organizations

U.S. Forest Service

U.S. Department of Interior

Wilderness Society

Web Sites

World Resources Institute
http://www.wri.org
Links: http://www.wri.org/wri/biodiv/foresthm.html

U.S. Department of the Interior
http://www.usgs.gov/doi/doi.html

Oregon Department of Forestry
http://salem10nt.odf.state.or.us/homepage.htm

Headwaters Forest
http://mercury.sfsu.edu/~brownb/

Endangered Habitats League
http://www.cyberg8t.com/wroberts/ehl/ehlwww.html

PROJECT NOTES

Tropical Forests ●

Tropical forests are among the most valuable and also among the most fragile resources on Earth; they help stabilize the global climate, purify air and water, support diverse human cultures, and sustain more than half of all animal and plant species on the planet. It is this biodiversity that contributes billions of dollars to the world's annual economy from the sale of commercial products derived from the forests' diverse plant life. These products include one-half of the global harvest of hardwoods, latex rubber, resins, essential oils, coffee, tea, cocoa, spices, and tropical fruits and at least one-quarter of the world's prescription drugs, representing about $100 billion of the yearly global economy. The National Cancer Institute (NCI) has isolated 2,100 plant species as sources of anticancer drugs and has barely scratched the surface of the tropical forest's vast array of animal and plant life that may harbor cures for many of our illnesses. In fact, despite the tremendous potential of finding useful products and pharmaceuticals, only about 1% of the estimated 125,000 species of flowering plants has been investigated. It is also anticipated that numerous tropical plants have food value or traits that could be transferred to food-crop plants to increase food production to help feed the soaring human population.

Despite the enormous importance of tropical forests to the global environment and to biodiversity, they represent only 7% of the world's land area. They are located in equatorial Asia, Africa, and Latin America, with over half concentrated in Brazil, Indonesia, Zaire, and Peru. There are several types of tropical forests depending on the patterns of rainfall. The tropical rain forests receive rainfall daily and are the ones that are being destroyed most rapidly. Tropical forests have survived for 150 million years, but today only about half remain; most of the damage has occurred since 1950. The most rapid decline occurred in the 1980s when about 9% of the world's tropical forests were lost; large numbers of small farmers slashed and

burned the forests in the Brazilian Amazon to establish small
farms and homes. In fact, these fires reached epidemic propor-
tions, causing both local and global air pollution. If current
trends continue unchecked, tropical forests may just about disappear by the year 2050.

IDEAS TO EXPLORE

Why are tropical forests ecologically vital to the global environment as well as to the global economy?

Why are they disappearing? How does rapid population growth
in nations that have tropical forests impact on these regions?
How have farming, logging, cattle ranching, tree plantations,
dams, and mining played a role in the decline of tropical forests?

Central America has given up much of its tropical forestland to
raise cattle for beef export. How does the high meat diet of the
rich nations of the world impact on the health of the tropical
forests?

Use graphs and charts to demonstrate the rates of decline in
Asia, Africa, and Latin America. How does this decline affect
global biodiversity? Show estimated trends.

PROJECT

How Do You Benefit from the Tropical Forests of the World? ● ○

Tropical forests benefit each of us daily through the products they provide as well as by the role they play in maintaining the atmosphere, climate, soil, water, and global
biodiversity. Find out how tropical forests touch your life.
List all of the products that you and your family use both
directly and indirectly. Think of the medicines you
benefited from over recent years and research their origin.
According to the Worldwatch Institute, over 80% of the

world's population gets its medicines from plants, especially from those that grow in the tropical forests. What wood products, food products, and materials that have enhanced your life are imported from tropical forests? What crops have been improved by importation of genes from tropical plants? According to the U.S. Department of Agriculture (USDA), genes bred to commercial crops have added $1 billion worth of food to the U.S. marketplace.

How do tropical forests contribute to the maintenance of the world's atmosphere, water resources, climate, and biodiversity? How do we all benefit from the ecological services of the tropical forests? Why is preserving biodiversity imperative for all of us? Think of why it is vital to keep wild species alive so that they can offer alternative forms of crops and products. Today, we depend on a very small number of crops like rice, wheat, soybeans, and maize, although there are numerous other edible species that exist. As water and soil resources dwindle, these other plants will become essential in feeding the world's exploding population. The rich reservoir of genes from the tropical forests is needed to help animals and plants survive changes in the Earth's climate. With so many diverse life forms, some will survive and adapt to environmental changes. By destroying the Earth's reservoir of genes, we are seriously impacting on our own long-term survival.

PROJECT

Select and Study a Tropical Forest Region ● ○

Tropical forests occur mostly in Asia, Africa, and Latin America. Although they stretch across more than 75 nations, over half of them are concentrated in four countries—Brazil, Peru, Zaire, and Indonesia. Select a tropical forest region or a nation with tropical forests. Use charts and diagrams to show changes in the tropical forests of your selected area over time and specify the different causes of the forests' decline. Are the forests being used for logging, ranching, and farms? What political, economic, and social factors are contributing to the destruction of the forests? What are some solutions? Costa Rica, for example, was almost entirely covered with tropical forests until politically powerful ranching families cleared much of the forests to graze cattle. Most of the produced beef was exported to the United States and Western Europe to help feed the industrial world's insatiable appetite for meat, leaving only about 17% of the original forest. Thus very few Costa Ricans benefited from the destruction of their tropical forests. In the mid-1970s, however, Costa Rica started to make a concerted effort to protect its rich reservoir of plants and animals by establishing a system of national parks and reserves along with extensive restoration projects. Numerous scientists, educators, and local residents are working to study and save the nation's biodiversity. As a result, Costa Rica has experienced a surge in tourism income; most of the tourists are interested in the country's great ecological resources (ecotourists).

RESOURCES

Published Materials

The Diversity of Life, 1992
    Edward O. Wilson
    W. W. Norton and Co., New York

Science, vol. 253, no. 5021, 1991, p. 760
    Paul R. Ehrlich and Edward O. Wilson

World Resources, A Guide to the Global Environment,
    1995–1996
    World Resources Institute in collaboration with the United
    Nations Environment Programme and United Nations
    Development Programme
    Oxford University Press, New York

Sustainable Harvest and Marketing of Rain Forest Products, 1992
    Edited by M. Plotkin and M. Famolare
    Island Press, Washington, D.C.

Tropical Timber Fact Sheet: The Woodbuyer's Guide
    Rainforest Action Network
    450 Sansome St., Suite 700
    San Francisco, CA 94111
    (415-398-4404)

Rain Forest in Your Kitchen, 1992
    M. Teitel
    Island Press, Washington, D.C.

Saving the Tropical Forests, 1988
    Judith Gradwohl and Russell Greenberg
    Earthscans Publications Ltd., London

Organizations

Rainforest Action Network

United Nations Food and Agriculture Organization

Web Sites

World Resources Institute
http://www.wri.org
Links: Forest Resources
http://www.wri.org/wri/biodiv/foresthm.html

Rainforest Action Network (RAN)
http://www.ran.org/ran/

United Nations Environment Programme (UNEP)
http://www.unep.ch/

Rainforest Workshop
http://mh.osd.wednet.edu

SYMBOL KEY

● Topic of average difficulty

△ Long-term assignment

\# Project requires special facilities, equipment, or supplies

○ Large public or college library required

∗ Safety precautions required

✦ Highly technical; specialized knowledge required

Preserving and Restoring the Tropical Forests of the World ● ○

According to many scientists, the most serious environmental problem that threatens our long-term survival is the world's rapid loss of its biodiversity. In order to stem the tide of tropical forest destruction and thus slow the depletion of our wild species, many changes need to be made on an international, national, regional, and local level. The Global Biodiversity Strategy developed by the United Nations Environment Programme (UNEP) describes conservation techniques that address the root causes of the forest devastation. These include ignorance about the value of products and medicines in the forests, government policies that subsidize farmers to clear land and farm unsustainable products as well as to form large-scale plantations to grow monocultures of products that are in global demand, failure to assess the value of biodiversity, population growth, and inefficient use of resources. It is becoming clear that in order to preserve and restore tropical forests integrative approaches that address these root causes need to be developed and implemented. This will require cooperative national and international policies along with regional and local inclusion and involvement at every step along the way.

IDEAS TO EXPLORE

What is meant by "Debt for nature swaps," an idea developed by the biologist Thomas Lovejoy? Has it worked?

Has nature tourism or ecotourism made a difference, especially in small countries like Costa Rica, Ecuador, and Kenya?

Are there instances of reserves created of tropical forests that are largely protected by local people who depend on them for their livelihoods? (Note Indonesia's Arfak Mountains Nature Reserve managed by the Hatam tribe.)

Are pharmaceutical companies involved in mining for medicines within the forests?

Should agricultural policies support growing indigenous crops that do not deplete water and soil resources? Give some examples.

How can we as individuals help encourage the sustainable use of tropical forests?

RESOURCES

Published Materials

Caring for the Earth, A Strategy for Sustainable Living, 1991
  Kreg Lindberg
  IUCN, Gland, Switzerland

*Policies for Maximizing Nature Tourism's Ecological and
  Economic Benefits,* 1991, p. 5
  World Resources Institute, Washington, D.C.

Science, vol. 239, 1988, p. 243
  Daniel H. Janzen

*Promoting Environmentally Sound Progress: What the North
  Can Do,* 1990, p. 5
  Robert Repetto
  World Resources Institute, Washington, D.C.

"U.S. Drug Firm Signs Up to Farm Tropical Forests"
  Washington Post, Sept. 21, 1991, p. A3
  William Booth

Organizations

World Conservation Union (IUCN)

United Nations Environment Programme

World Wildlife Fund

Web Sites

**Worldwide Rainforest/Biodiversity Campaign News
Internet Archives**
gopher://gaia1.ies.wisc.edu:70/00/research/wforests/

Tropical Biodiversity
http://www.bdt.org.br/bioline/bin/tb.cgi

Tropical Forests
http://wri.org/wri/biodiv/tropical.html
Links: Saving Tropical Forests
http://www.wri.org/wri/biodiv/tol-fore.html

El Planetor Platica: Eco Travels in Latin America
gopher://csf.Colorado.edu/oo/environment/orgs/El_Planeta_
Platica
http://www.planeta.com/

SYMBOL KEY

●   Topic of average difficulty

△   Long-term assignment

\#   Project requires special facilities, equipment, or supplies

○   Large public or college library required

*   Safety precautions required

✚   Highly technical; specialized knowledge required

PROJECT NOTES

Solid Waste

Any solid material that is discarded constitutes solid waste. Although only 5% of the world's population resides in the United States, we contribute about one-third of the volume of total global solid waste. Each of us discards about 1,500 lb. per year of solid waste, most of which ends up in landfills or is incinerated. We presently spend billions of dollars annually to discard solid wastes, and future costs are predicted to skyrocket. Furthermore, landfills pollute land and water resources while incineration adds to air pollution. In fact, incineration produces dioxins which are highly carcinogenic contaminants. Clearly, part of the solution must be to decrease the amount of solid waste discarded by each of us.

PROJECT

Reducing Solid Waste in Your Community ●

For a selected period (one week, for example) record the amount and nature of the solid waste your family discards. What could be recycled or reused for some other purpose? Could brands that contain recyclable materials be substituted? Could you buy more durable items or buy in bulk to reduce packaging? Analyze by how much your family could reduce solid waste. Based on your observations and calculations, estimate the potential waste reduction of the community if all families were able to reduce waste by the same percentage as your family. Also, suggest how your community could decrease solid waste by using yard waste for compost, for example.

RESOURCES

Published Materials

Decision-makers Guide to Solid Waste Management, Nov. 1989
    U.S. Environmental Protection Agency

The Waste System, 1988
    U.S. Environmental Protection Agency

Design for a Livable Planet, 1990
    Harper and Row, New York

Web Sites

Start with the general environmental web sites listed in Part III
of this book.

SYMBOL KEY

● Topic of average difficulty

△ Long-term assignment

\# Project requires special facilities, equipment, or
  supplies

○ Large public or college library required

* Safety precautions required

✚ Highly technical; specialized knowledge required

Internet Projects ● △

Select an EPA press release about pollution released into the environment and track the impact of both the press release and the pollution itself and the clean-up efforts. Addresses, telephone numbers, and e-mail addresses are generally provided for the agency and people involved in the process. Express your views on the efficiency of the process and what might be done in order to avoid recurrence of the problem.

EPA press releases
http://www.epa.gov./docs/PressReleases/

Select a pesticide that has just been taken off the market and trace its path from beginning to end, determining how it was originally approved and why it is now banned. Express your views about the approval process and how toxic chemicals can still be spread into the environment.

U.S. Environmental Protection Agency
http://www.epa.gov

Pesticide Action Network North America Update Service (PANUPS)
gopher://gopher.igc.apc.org.:70/11/orgs/panna/panups/
panups_pointer

Follow the monarch butterfly's fall migration. Use the Internet to note sightings along the way. Compare to past migrations.

Monarch Watch
http://129.237.246.134/

Terrestrial Biomes ●

A biome is a region that supports a characteristic, naturally occurring community of plants and animals. Each biome is also characterized by specific environmental conditions. The major biomes on land are deserts, grasslands, and forests, with subdivisions within each.

The kind of biome that predominates in any particular region is determined largely by the average annual precipitation and temperature. The average annual precipitation in a region will generally determine whether it is a desert, grassland, or forest, while the average annual temperature and soil composition will determine if a biome is tropical, temperate, or polar. Deserts usually receive less than 2.5 cm (10 in.) of precipitation annually; grasslands have sufficient quantities of precipitation to support grasses and maybe a few trees, but droughts and fires do not permit large tree stands; moderate to large quantities of precipitation fall on the forested regions of the Earth. Natural, undisturbed forests consist of various species of trees as well as many other forms of smaller plant life.

IDEAS TO EXPLORE

Select one from each of the several classifications of terrestrial biomes and describe them. Where are they located? What kinds of plant life do they support? What animals can therefore survive?

What role does each of these biomes play in maintaining the environment? Do they prevent soil loss during storms? Do they prevent flooding? support important wildlife? purify water and air? stabilize the world's climate? recharge groundwater?

Which of these biomes in particular are being degraded by human activities like logging, grazing, and farming?

PROJECT

Which Biome Do You Live In? ● ○

Identify the biome that characterizes your community or state. What plants and animals lived in the region before the arrival of Europeans? What changes have occurred and how have they impacted on the plant and animal life? What remains today? Have foreign species been introduced? Has any of the original community of plants and animals been preserved? What has been the impact of human activities on the soil, water, and air of your region? Use charts and graphs where possible to show changes over time.

RESOURCES

Published Materials

Earth, 1992
   Derek Elsom
   Macmillan Publishing, New York

Refer to a college-level environmental science or biology text book to read about biomes. Contact your local environmental agencies, universities, and organizations.

Web Sites

List of State Environmental Agencies
http://www.tribnet.com/environ/env_stat.htm

List of University Environmental Web Sites
http://bigmac.civil.mtu.edu/aeep/univ.html

Public Lands in the U.S. ● ○

The United States leads the world in reserving land for public use, recreation, and wildlife. About one-third of our nation's land belongs to all of us: 73% of this is in Alaska and about 22% in the West. In fact, about 60% of the West is considered public lands. Much of the land in these regions is used for logging, mining, livestock grazing, and oil and gas extraction; other public lands have more restricted uses. The National Park System, for example, includes 52 major national parks that can only be used for hiking, camping, fishing, and boating. Recently there has been growing pressure by industry to make more public lands available for gas, oil, and mineral extraction, as well as for logging and grazing. Alaska, which contains much land rich in oil and mineral reserves, is eager to open public land to industry in order to boost the local economy, as are a number of other states.

IDEAS TO EXPLORE

How are our public lands classified? Who is using them and at what cost?

How does the federal government subsidize ranchers and timber, oil, and mining companies? Who pays for the environmental degradation resulting from these activities?

How should public lands be managed so that they enhance economic growth in a sustainable fashion?

PROJECT

Focus on the Environmental History of a Major National or State Park ● ○

Focus on a national park or other form of public land and trace its history. Why and how was it protected? Have human activities impacted on the health of the park? Have surrounding agricultural activities depleted water resources for the park (as is the case with the Everglades in Florida)? Is air pollution destroying forests (as in Smoky Mountain National Park), decreasing visibility (as in the Grand Canyon), or acidifying lakes and ponds (as in Adirondack State Park in New York)? What has been the impact of hikers on vegetation and wildlife? Are there growing problems with wildlife-human contact as more and more people visit the parks?

PROJECT

Identify Public Lands in Your State and Examine How They Are Being Managed ● ○

Construct a topographical map of your state indicating the locations of all local, state, and federal public lands. Find out which restrictions apply to their use and which agencies are responsible for their protection. How have they been used? Are they being adequately protected? Are they being degraded? Develop a plan to improve the management of these public areas.

RESOURCES

Published Materials

The Amicus Journal (published by the Natural Resources Defense Council), Summer 1996, p. 26
K. Durbin

National Parks (the magazine of the National Parks and Conservation Association), Sept./Oct. 1993, p. 26
R. Stapleton

Playing God in Yellowstone: The Destruction of America's First National Park, 1987
Alston Chase
Harcourt Brace and Company, New York

Organizations

U.S. National Park Service

U.S. Fish and Wildlife Service

U.S. Forest Service

U.S. Department of the Interior

Nature Conservancy

National Park and Conservation Association (NPCA)

Web Sites

Environmental Organization Web Directory
http://www.webdirectory.com/

U.S. Fish and Wildlife Service
http://www.fws.gov/

List of State Environmental Agencies
http://www.tribnet.com/environ/env_stat.htm

The Impact of Acid Rain on American Forests ● ○

For decades midwestern power plants have released large quantities of sulfur and nitrogen oxides into the atmosphere where prevailing wind currents carry them over the northeastern United States and eastern Canada. There they combine with precipitation to form acid rain. A recent long-term study conducted over several decades on a forest stand in New England concluded that acid rain is depleting forest soil of essential nutrients like calcium. The low concentration of calcium is associated with depressed plant growth; in fact, the forest studied showed zero growth nearly a decade ago and is only slowly recovering. Scientists are concerned that currently healthy forests in these areas may one day decline as soil nutrients continue to be depleted by acid rain events. Interpreters of the study also claim that acid rain is responsible for the decline of high-altitude red spruce trees and sugar maples in eastern Canada and that it is harming aquatic life in about one-tenth of eastern lakes and streams.

IDEAS TO EXPLORE

Which forests and types of trees in the East have been reported as being in decline? How are they damaged?

Why has it been difficult to prove scientifically that acid rain is the culprit?

PROJECT

Select and Study a Specific Forest ● ○

Select a forest in your state or in another region of the United States. Use local and state agencies, organizations, and universities to determine the composition and health

(continues)

of the forest. Find acid rain data in local reports by environmental agencies, organizations, or universities. (Acid rain data is often provided on the Internet.) Has the forest been subjected to acid rain? Have some species of trees declined? Use charts, maps, and tables. Has the pH of the soil changed?

RESOURCES

Published Materials

Science, Apr. 1996, p. 116
G. E. Likens et al.

"Northeastern Forest Soil Impacted,"
New York Times News Service, Apr. 16, 1996
William K. Stevens

Ill Winds: Airborne Pollution's Toll on Trees and Crops, 1988
James J. MacKenzie and Mohammed T. El-Ashry
World Resources Institute, Washington, D.C.

Interim Assessment: The Causes and Effects of Acid Deposition, vol. 4
National Acid Precipitation Assessment Program
U.S. Government Printing Office, Washington, D.C.

Organizations

U.S. Department of the Interior

U.S. Environmental Protection Institute

Web Sites

Forest Resources
http://www.wri.org/wri/biodiv/foresthm.html

U.S. Department of the Interior
http://www.usgs.gov/doi/doi.html

Forest Service Ecosystem Management
http://www.fs.fed.us/land/welcome.htm

Gaia Forest Archives
http://gaia1.ies.wisc.edu/research/pngfores/gaia.html

SYMBOL KEY

● Topic of average difficulty

△ Long-term assignment

\# Project requires special facilities, equipment, or supplies

○ Large public or college library required

* Safety precautions required

✚ Highly technical; specialized knowledge required

The Psychology of Recycling ● ○

Recycling is the most environmentally preferred way of managing our solid wastes. Incineration and landfills are the main other methods of solid waste disposal, but they either directly or indirectly cause pollution of the land, water, and air. Recycling also conserves resources. Although most Americans want to protect the environment, many do not recycle until mandatory recycling is instituted. In fact, recycling has been mandated in ten states as a result of a poor response to voluntary programs. Low recycling rates, for example, prompted Hamburg, New York, to implement a mandatory recycling program: trash brought to landfills has been reduced by 34% and thousands of dollars have been saved in disposal costs.

In a time when resources are dwindling and billions of tons of solid wastes are being generated annually, it is imperative that Americans and American industries be motivated to conserve, reuse, and recycle.

PROJECT

The Psychology of Recycling ● ○

Interview members of your community about their attitudes on recycling. Find out why some people recycle and why others do not. What reasons do people give for not participating in a recycling program? What do they think would be needed to motivate them to recycle? Do they understand the solid waste problem and its impact on the environment? Is more education needed? What medium would be most effective in delivering the message? Are incentives to recycle more effective than punishments? Investigate some of the voluntary and mandatory programs now in operation. Which ones are successful? What do they appear to have in common that contributes to their success? On the basis of your research, develop a recycling program for your community to improve recycling rates.

RESOURCES

Published Materials

The McGraw-Hill Recycling Handbook, 1993
  Herbert F. Lund
  McGraw-Hill, Inc., New York

Environmental Problems/Behavioral Solutions, 1980
  John D. Cone
  Brooks/Cole Publishing, California

Public Relations Review, vol. 10, no. 4, Winter 1984, p. 23
  Mark A. Larson and Karen L. Massetti-Miller

Environment and Behavior, vol. 11, 1979, p. 539
  P. D. Luyben and J. S. Bailey

Journal of Applied Social Psychology, vol. 16, no. 1, 1986, p. 29
  Shawn M. Burn and Stuart Oskamp

Web Sites

Texas Natural Resource Conservation Commission (TNRCC)
http://www.tnrcc.state.tx.us/

The EcoWeb, University of Virginia
http://ecosys.dr.dr.virginia.edu/EcoWeb.html

Environmental Defense Fund
http://www.sun-angel.com.edf/cdf.html

Sustainable Agriculture ● ○

The growth of food output is slowing, largely as a result of environmental degradation. Loss of soil through erosion, declining water resources for irrigation, air pollution, ozone depletion (with more UV light reaching plants), and hotter summers (perhaps from global warming) are all slowing crop production. For example, it has been estimated that farmers worldwide are losing about 24 billion tons of topsoil annually. This is equivalent to the loss of about an inch of topsoil from about one-half of China's cropland. With the global population increasing by 93 million per year, we need to grow more food, not less. Clearly, our precious soil must be preserved to avoid increased famine and the resultant political and social upheaval. Lester Brown, the president of Worldwatch Institute, calls our soil depletion a "Quiet Crisis" the effect of which will be more catastrophic than oil depletion. To keep the world's fields fertile, agricultural practices must radically change from those that degrade soil and water to those that sustain the environment and maintain the fertility of the soil.

IDEAS TO EXPLORE

How do conventional agricultural practices cause soil erosion? How do tilling and the use of synthetic fertilizers and pesticides degrade the soil and water?

How do pesticides impact on wildlife? Why is there growing concern today that some pesticides may be very harmful to human health?

What are some agricultural practices that sustain and enhance soil and water resources? What is meant by alternative agriculture, and where is it being practiced?

PROJECT

Replacing Pesticides with Integrated Pest Management ● ○

Integrated Pest Management (IPM) is an agricultural practice that uses biological pest control, crop rotation, and timing of planting along with limited pesticide usage where absolutely necessary to minimize damage to crops. The use of few or no commercial pesticides reduces damage to the environment and expense to the farmer. It is widely recognized that many chemical pesticides, herbicides, and fungicides are polluting natural water systems, harming ecosystems, and decreasing biodiversity. In addition, many scientists are concerned about the impact of the hundreds of different pesticides on human health. For this project, select a farm in your state that relies heavily on pesticides and develop an IPM plan that could replace the heavy use of chemical pesticides. Perhaps you can locate a farm through your local agricultural extension service or a school of agriculture at a state university. In developing your strategy, find out about the major pests and weeds that threaten the crops. Below are some of the questions you should explore in researching your plan:

- What are the natural predators to the pests? Are these predators also damaging to the crops?
- Could other crops that attract these natural predators be planted nearby?
- Are there times during the season when pheromones can be sprayed to decrease proliferation of the pests? Can planting be timed to take advantage of times when pest populations are low?

If possible, interview the farmer(s) to find out about their experiences with other approaches to controlling pests.

PROJECT

Experimenting with Natural Herbicides and Pesticides ●

There are numerous natural herbicides and pesticides that gardeners and farmers have used since the advent of farming 10,000 years ago. Research several of these to identify plants that you can grow on a small plot of land. Marigold plants, for example, are often planted in vegetable gardens because they release chemicals that repel pests that damage vegetable plants. Research the literature and ask local nurseries, farmers, and gardeners as well as county agricultural extension services for information about these natural chemicals. Select several and then prepare your own natural herbicide or pesticide solution and test it on an experimental garden. You will also need a control plot that you do not treat in any way.

PROJECT

Design a Farm Using Alternative Agriculture ● ○

Alternative agriculture refers to farming practices that reduce harmful effects on the environment, preserve water and soil resources, protect human and animal health, and reduce costs and raise efficiency for the farmer. They constitute a wide range of practices including IPM, rotation of crops, little or no dependence on chemicals for fertilizers and pest control, little or no tilling of soil, etc. Design a farm that could exist in your state. Perhaps you can select the crops and livestock that are presently raised in your state or county. Research the various alternative techniques that have been used and apply those you think are suitable to design your farm.

PROJECT

The Impact of Pesticides on Soil Organisms ●

Soil constitutes a very complex ecosystem. Each handful of soil contains millions of organisms most of which are too small to see except with a microscope (fungi and bacteria) or hand lens (nematodes and mites). Other organisms, like ants and worms, can be readily identified. All of these species play a role in maintaining the fertility of the soil. Chemical pesticides can actually be harmful to soil because they destroy other organisms in addition to the target species. For this project, obtain from a farm, garden, or perhaps a golf course soil that has been routinely sprayed with chemical pesticides. Select an organism that can be seen with a magnifying glass or microscope that is prevalent in healthy soil. Count the numbers of this organism in several samples of soil contaminated with pesticides and then compare with numbers counted in soil that has not been exposed to pesticides. Be sure to avoid any direct contact with pesticide-contaminated soil.

Hundreds of thousands of tons of chemical pesticides are sprayed onto farmland in the United States annually. Often, however, crop yields are not increased, and sometimes they are actually lessened. One reason is that the pesticides often kill predator organisms—organisms that keep down the population of plant-eating species. The resulting population explosion of plant-eating organisms proceeds to demolish the crop plants. Also, heavy use of pesticides in the United States has significantly reduced the honeybee and wild bee populations that are needed to pollinate about one-third of our fruits, vegetables, and forage crops. It is interesting to note that before the heavy use of synthetic fertilizers and pesticides, farmers lost about 20% of their crops as compared with the same 20% today.

RESOURCES

Published Materials

Alternative Agriculture, 1989
   National Research Counsel
   National Academy Press, Washington, D.C.

Saving Our Soil, 1995
   J. Glanz
   Johnson Books, Boulder, Colo.

Silent Spring Revisited, 1987
   Edited by G. Marco et al.
   American Chemical Society, Washington, D.C.

The Pesticide Conspiracy, 1980
   R. Van den Bosh
   Anchor Books, Garden City, N.Y.

Saving the Planet, 1991
   Lester R. Brown et al.
   W. W. Norton and Co., New York

Ecology of Pesticides, 1978
   W. A. Brown
   Wiley, New York

American Journal of Alternative Agriculture,
   vol. 2, no. 3, 1987, p. 99
   J. W. Doran et al.

American Journal of Alternative Agriculture,
   vol. 2, no. 1, 1987, p. 3
   G. R. Hallberg

Horticultural Science, vol. 21, 1986, p. 397
   S. D. Holmberg et al.

Agricultural Soil Loss: Processes, Policies, and Prospects, 1987
   J. M. Harlin and A. Hawkins
   Westview Press, Boulder, Colo.

Organizations

U.S. Department of Agriculture (USDA)

Web Sites

Alternative Agriculture News
http://envirolink.org/pubs/Alternative_Ag_News

Earth Negotiations Bulletin
http://www.iisd.ca/linkages/voltoc.html

Sustainable Development Department of the Food and Agricultural Organization of the United Nations
http://www.fao.org/waicent/faoinfo/sustdev/welcome_.htm

SYMBOL KEY

● Topic of average difficulty

△ Long-term assignment

\# Project requires special facilities, equipment, or supplies

○ Large public or college library required

✻ Safety precautions required

✚ Highly technical; specialized knowledge required

PROJECT NOTES

Fertilization Techniques

The humus or organic content of the soil is vital to its fertility. Humus retains moisture, insulates the soil from extreme temperatures, reduces evaporation, and keeps nutrients in the soil, later releasing them in forms that are easily absorbed by the plants. Soils rich in humus are also better able to resist flooding by holding moisture during precipitation events. And humus-rich soils filter pollutants out of water thereby maintaining the quality of our water resources.

PROJECT

How Fertilization Techniques Determine the Organic Content of Soil ●

To study the impact of synthetic versus natural fertilizers on the organic content of soil, locate soil that has been periodically fertilized with synthetic fertilizers only. Remove a small sample—about half a cup. Also locate soil that has been replenished with natural fertilizers like compost or manure. (Talk with local gardeners to find these samples.) Remove about the same amount of soil from each source. For each sample taken, carefully remove all litter—leaves and sticks and large pebbles and stones. Crush the soil into small particles. Use aluminum foil to construct small cups that can hold a small amount of each sample. Weigh each cup and then pour in enough soil to cover the bottom of the cup. Weigh each cup now filled with soil. You can use a food scale at home or perhaps a scale in a science lab at your school. By subtracting the weight of the empty cup from the weight of the filled cup, you can determine the weight of each sample. To find the percentage of organic content of each sample, place the

(continues)

cups containing the soil on a baking pan and let them dry out by evaporation for several days to remove the moisture in the soil. Weigh them again. Next, bake the soil in an oven, preferably one in a science lab at school. Alternatively, you can use your home oven, setting it at a high temperature for about 10 to 15 minutes. (Make sure a fan is on in case of smoking.) Let them cool and then reweigh. The organic component of each sample of soil will be burned off. The organic component is the amount by which your final weight differs from your weight after drying and before heating. Using your weights, you can now calculate the percentage of organic material for each sample. (You can also calculate the percentage of moisture in each sample.)

SYMBOL KEY

● Topic of average difficulty

△ Long-term assignment

\# Project requires special facilities, equipment, or supplies

○ Large public or college library required

✱ Safety precautions required

✚ Highly technical; specialized knowledge required

SAMPLE	CUP WT	SOIL + CUP WT	SOIL WT	CUP + DRY SOIL WT	CUP + BAKED SOIL WT	ORGANIC WT %
1						
2						
3						
4						
5						
6						
7						
8						
9						
10						

You can use the following formulas:

Dry Soil wt   = (Cup + Dry soil) – Cup
Organic wt   = (Cup + Dry soil) – (Cup + Baked Soil wt)
% organic material = (Organic wt ÷ Soil wt) × 100
% moisture   = (Soil + Cup wt) – (Dry Soil + Cup wt) ÷ (Soil wt)

Remember: the larger the number of samples taken the more valid will be your conclusions. To make your results more reliable you should take several samples of soil at different locations from the same garden.

RESOURCES

Published Materials

The Soil Resource, 1979
Hans Jenny
Springer Verlag, New York

Soil Management for Sustainability, 1991
Edited by R. Lal and F. J. Pierce
Soil and Water Conservation Society, World Association of
Soil and Water Conservation and Soil Science Society of
America

Organizations

U.S. Department of Agriculture

SYMBOL KEY

● Topic of average difficulty

△ Long-term assignment

Project requires special facilities, equipment, or supplies

○ Large public or college library required

＊ Safety precautions required

✚ Highly technical; specialized knowledge required

The Role of Humus in the Soil

Humus is a complex mixture of organic substances that are formed whenever plant or animal matter decomposes sufficiently so that it can no longer be recognized. Humus is a critical component of soil in that it retains moisture, holds nutrients in the soil, and releases nutrients to plants in a form that can be readily absorbed. Humus is the habitat for millions of organisms that are necessary for the maintenance of soil fertility.

PROJECT

The Interaction of Water with Humus ●

Obtain samples of rich soil from a forest or garden as well as samples of poor soil from an eroded surface or from some other source. Place measured samples on a napkin, paper towel, or some other porous paper that you can insert into a filter. Pour measured amounts of water onto the samples and catch any water that filters through. Which of the samples retains the most water? How long does the water take to filter through? Why is it especially important to have rich soils in regions with minimal rainfall? How do forest soils prevent flooding? What happens during periods of heavy rainfall in regions that are largely deforested where the soil has eroded and has a low humus content as a result?

PROJECT

The Filtering Action of Humus in the Soil ●

Use several samples each of rich soil and soil low in humus content. Place measured amounts of each in a porous paper product that lines the inside of a filter. Pour water with a known amount of food coloring through each sample. What is the color of the effluent? Has it retained the color of the original solution or has most of it been filtered out by some of the samples? (Fertile soil not only provides us with crop plants but also filters our drinking water.) Food dye, like many water pollutants, is an organic substance. Just as the dye is retained by fertile soil, so, too, are harmful water contaminants.

PROJECT

Humus and Rates of Decay ● △

Millions of organisms interact with the humus component of soil. Many of these feed on the plant and animal matter that falls on the soil's surface decomposing organic matter in the process. Rich soils can decompose more organic matter and at a faster rate. Select several soil sites with varying quantities of humus. Bury paper products, kitchen wastes, and leaves in different sites. Check on them periodically to determine their current state? Which decomposes faster and in which soils? Try placing them at different depths? Does the soil become less fertile as you dig deeper? Why?

RESOURCES

Published Materials

Nature, vol. 371, Oct. 1994, p. 783
    H. Tiessen et al.

Sun, Soil, and Survival: An Introduction to Soils, 1978
    K. C. Berger
    University of Oklahoma Press, Norman, Okla.

Organizations

U.S. Department of the Interior

U.S. Department of Agriculture

SYMBOL KEY

●   Topic of average difficulty

△   Long-term assignment

Project requires special facilities, equipment, or supplies

○   Large public or college library required

＊   Safety precautions required

✚   Highly technical; specialized knowledge required

PROJECT NOTES

The Leaching of Mineral Nutrients From Soil ● ○

Plants require about 13 different mineral nutrients to grow. Three of these—nitrogen, phosphorus, and potassium—are particularly important because they are absorbed by plants in significant quantities. Consequently, as plants mature they deplete the soil of these nutrients. Water from precipitation or irrigation also removes nutrients from soil as it seeps into the ground by dissolving nutrients and carrying them to deeper layers in a process called **leaching.** In the natural world, soil nutrients are replaced by the disintegration of rocks and the decomposition of animal and plant matter. However, when we repeatedly plant and water crops on farmland, soil nutrients are removed at a much more rapid rate than that at which they can be naturally restored. In order to keep growing crops on the same land most farmers and gardeners add commercial synthetic fertilizers. These artificial fertilizers contain primarily nitrogen as ammonium ions, nitrate ions or urea, phosphorus as phosphate ions, and potassium as potassium ions. Unfortunately, the N, P, and K in these chemical forms tend to readily leach out of the soil thereby contributing to the pollution of groundwater and of lakes, rivers, streams, bays, and estuaries. In fact, many wells in farm areas are polluted with high levels of nitrates that may be toxic to infants. Also, the exclusive use of chemical fertilizers reduces the organic matter in the soil so that the soil's capacity to hold both the water and oxygen that are needed for plant growth declines and the soil becomes less capable of supporting crops. This physical degradation of soil is a cause of soil fertility loss in the United States and in other industrialized nations. Natural fertilizers that come from animal manure or decomposed plant matter are an alternative to synthetic or inorganic fertilizers. These organic fertilizers not only restore nutrients to the soil but also enhance its water- and oxygen-carrying capacity. Natural fertilizers lose their nutrients slowly and do not pollute water or degrade soils. It is especially important for suburban residents to rely on organic fertilizers for the maintenance of lawns. The widespread use of synthetic fertilizers in suburban communities in particular has contributed significantly to the deterioration of rivers and streams.

IDEAS TO EXPLORE

How much artificial fertilizer do American farmers use? What happens to much of these chemicals during periods of heavy rainfall? at other times?

What is the impact of these nutrients on the health of streams, rivers, and lakes? For example, do they cause the eutrophication of lakes? How do they impact on the indigenous fish?

Do they end up polluting drinking water? What are their effects on human health?

Are there alternatives to fertilizing the soil with such large quantities of these chemicals? Is there a difference between the nutritional value of crop plants exposed to artificial fertilizers and those naturally fertilized?

PROJECT

The Leaching of Soil Nutrients from Synthetic and Natural Fertilizers ● # *

The purpose of this project is to compare the rate of leaching of plant nutrients from soil mixed with natural fertilizers versus soil mixed with synthetic fertilizers. Obtain a soil-testing kit from a local nursery or garden supply store. (Soil-testing kits are readily available in most stores that sell garden or farm supplies.) Take several (four or more) soil samples from the same area in your neighborhood and use the test kits to determine the levels of nitrogen, phosphorus, and potassium in the samples. The levels should be very similar from sample to sample. Then obtain commercial synthetic fertilizer and natural fertilizers (animal manure, decomposed plant materials, or compost). You can generally obtain both types of fertilizer from a garden

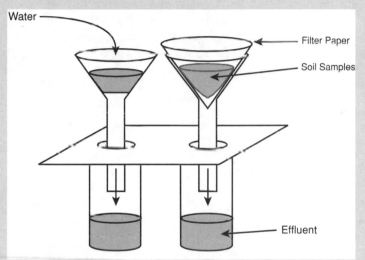

supply store. Mix the fertilizers into the samples so that two or more soil samples contain the inorganic while the others contain organic fertilizer. (The more samples you use, the more reliable will be your test results.) Determine the appropriate amounts to mix from the directions on the packages and follow all safety precautions with regard to proper handling of the fertilizers. After mixing, use your test kit to measure the nutrient content of each sample. Transfer the soil samples into funnels lined with filters. (Coffee filters or paper towels can be used as filters.) Bore two holes into a plywood board for two funnels and place the board on two containers (pails or plastic milk containers) that will be used to catch the filtered water. Insert each funnel into a hole and fill each with a soil sample as shown in the diagram.

(continues)

Pour the same quantity of water through each sample. Perhaps you could approximate the amount of water that would fall on the samples in one or more growing seasons if the soils were to be used to raise a specific crop. Some crops, like cotton and alfalfa, are water intensive while others require less irrigation. Measure the nutrient content of the samples after exposure to water and compare. Take three measurements for each sample.

Your test results should answer the following:

Which sample lost more of its nutrients? How much was lost? Do the nutrients leach out at the same rate? Try testing the water in each of the receptacles; if the water samples are too dilute, heat them gently to drive off some of the water and concentrate the nutrients. You could use your results to estimate the amount of the three nutrients that would leach into groundwater from a field of planted crops.

RESOURCES

Published Materials

World Resources, A Guide to the Global Environment, 1994–1995
World Resources Institute
Oxford University Press, New York

Scientific American, June 1990, p. 112
J. Reganold et al.

Organizations

U.S. Dept. of Agriculture and your state and local extension

American Farmland Trust

Web Sites

Food and Agricultural Organization of the United Nations
Sustainable Development Department
http://www.fao.org/waicent/faoinfo/sustdev/welcome_.html

Great Lakes National Program Office Homepage
http://www.epa.gov/glnpo

Agricultural Bibliographic Information System of the Netherlands
http://www.bib.wau.nl/agralin/opac.html

SYMBOL KEY

● Topic of average difficulty

△ Long-term assignment

\# Project requires special facilities, equipment, or supplies

○ Large public or college library required

* Safety precautions required

✚ Highly technical; specialized knowledge required

Attracting Earthworms for Healthier Soil ●

Soil scientists are giving increasing recognition to the critical role played by numerous and varied life forms in forming and maintaining the soil. Soil was once viewed as being a result of physical, chemical, and biological forces that acted predictably over time. In current thinking, soil is emerging as a sea of life, a spongy home to billions of organisms that interact with each other, with organic matter, and with rocks and minerals in complex ways. Earthworms in particular are growing in value to farmers who are seeking to lessen soil erosion and to maintain and build the organic content of the fields so as to ensure long-term productivity.

Worms are burrowing animals that create miles of tunnels bringing essential air and water to the inner levels of the soil. Like other burrowing life forms, they help mix the decaying organic matter from the surface into the deeper layers of land by ingesting plant residue and excreting at other sites. Earthworms generally eat about four times their body weight daily, so their impact on moving organic matter from plant litter into the soil is considerable. Earthworms have been thought of as a natural tilling organism, providing the benefits of tilling without the negative side effects. Equally important is the impact of millions of tunnels on the surface. The tunnel holes act as sponges to soak up the rain, minimizing erosion by water as well as helping to stop flooding during heavy rainfall. Farms that do not till their soil, thereby maintaining high concentrations of earthworms, lose far less of their soil during rainstorms than do conventional farms.

PROJECT

How to Attract Earthworms to Your Farm or Garden ● △

Designate two or more small plots of land for your study. Use different types of organic matter to attract worms. For example, you can use grass clippings, pine needles, pine bark, or leaves. You can mix them or try them in layers. Use various combinations to determine if one is more successful than another in drawing earthworms to your experimental plots. You can also try mixing in animal manure. You might count earthworms that you uncover in a specified amount of soil or perhaps count tunnels and measure their lengths.

PROJECT

The Impact of Tilling on Earthworms ●

Soil that is tilled generally contains a lower population of earthworms. Tilling the soil causes rapid consumption of plant residue by soil microbes, leaving less for earthworms to eat. Count earthworms or their tunnels in both tilled and untilled soil.

RESOURCES

Published Materials

Geoderma, vol. 46, 1990, p. 73
    W. M. Edwards et al.

Saving Our Soil, 1995
J. Glanz
Johnson Books, Boulder, Colo.

National Conservation Tillage Digest, Feb. 1994, p. 16
E. Kladiviko

Organizations

Dept. of Plant and Soil Science, University of Tennessee
Institute of Agriculture

Web Sites

Alternative Agriculture News
http://envirolink.org/pubs/Alternative_Ag_News

Soil, Water, and Climate
http://www.soils.umn.edu/

Farmer to Farmer
http://www.organic.com/Non.profits/F2F/

United States Department of Agriculture
http://www.usda.gov/usda.htm

The Living Soil ● ○

Our view of soil is changing significantly as more and more studies show the immense numbers and types of organisms that live in the soil and contribute to its fertility. In one teaspoon of healthy soil there are about 100 million individual bacteria, a mile of fungal hyphae, thousands of algal and protozoan cells, and hundreds of nematodes. Along with these microscopic forms, burrowing animals like worms, nightcrawlers, ants, termites, crayfish, wasps, millipedes, beetles, pill bugs, and even larger organisms like gophers, mole rats, prairie dogs, and many others contribute to the health of the soil. The interaction of these life forms with the components of the soil and with each other produces a dynamic view of soil called **bioperturbation.**

IDEAS TO EXPLORE

Why is soil really a complex ecosystem? What do we know about some of the processes and interactions that maintain the soil?

What are the effects of conventional farming on the health of the soil? What is the impact of the use of pesticides, synthetic fertilizers, tilling, heavy farm machinery, and irrigation methods on the soil?

How do deforestation and overgrazing impact on the health of soil?

What can farmers do to conserve their soil and to ensure the productivity of their farms for the future?

PROJECT

The Impact of Pesticides on Soil Organisms ● *

One of the many problems with pesticide applications is that in addition to destroying the target pests the pesticides kill or harm numerous other organisms, many of

(continues)

which are beneficial to the soil and to crop plants. In some instances pesticides have actually caused more damage to plants because they decimated populations of predator insects, causing the population of plant-eating insects to explode as a result. In addition, pesticides collect in streams, rivers, lakes, and coastal regions, polluting drinking water and often harming wildlife (birds, for example). There is much concern today that many pesticides may be harmful to human health, especially that of children.

Obtain several samples of soil from farmland or from a garden that has been sprayed with pesticides. Compare with several samples of soil from a farm or garden where pesticides are not used. Can you identify differences in the populations of organisms living in the samples? Which pesticides were used? Are some more harmful than others to soil life forms?

PROJECT

The Impact of Tilling on Soil ●

Over the decades, the tractor has become almost the symbol of the American farmer, but today there is much controversy over the impact of tilling on long-term soil fertility. Unfortunately, there is growing evidence that the constant turning over of the farmland has caused degradation of the soil. When soil is tilled, plants that secure that soil to the surface of the land are uprooted so that wind and water are able to remove the topsoil (erosion). Turning the soil over infuses the earth with oxygen and buries the plant matter. Tiny organisms in the soil rapidly consume the buried organic material. As a result there is less organic matter available for the following planting season. In general, farmland that is constantly subjected to tilling is lower in humus content and supports a much smaller population of earthworms. In addition, heavy farm machinery causes compaction of the soil and decreases its ability to absorb water and air. Many farmers today are using a no-till approach to raising their crops and are preserving their soil.

Examine soil samples from farms where tilling is routinely used and compare with soil samples where a no-till approach has been adopted. Compare the amounts and kinds of organisms that you can find. You can also determine the organic content of the samples by using the method described in the project "How Fertilization Techniques Determine the Organic Content of the Soil" (earlier in this chapter).

RESOURCES

Published Materials

Soil Science, vol. 149, 1990, p. 84
    D. L. Johnson

Saving Our Soil, 1995
    James Glanz
    Johnson Books, Boulder, Colo.

Agronomy Journal, vol. 85, no. 6, 1993, p. 1237
    D. C. Reicosky et al.

Management to Reduce Erosion and Improve Soil Quality
    Appalachia and Northeast Region, USDA Agricultural
    Research Service Conservation Research Report No. 40,
    1995

Farm Journal, Apr. 1993, p. 19
    S. Hillgren and D. Smith

Organizations

U.S. Department of Agriculture (USDA)

Web Sites

U.S. Department of Agriculture
http://www.usda.gov/usda.htm

Soil, Water, and Climate
http://www.soils.umn.edu/

Sustainable Development Department (SDD) of the United Nations Food and Agriculture Organization
http://www.fao.org/waicent/faoinfo/sustdev/welcome_.htm

SYMBOL KEY

● Topic of average difficulty

△ Long-term assignment

Project requires special facilities, equipment, or supplies

○ Large public or college library required

* Safety precautions required

+ Highly technical; specialized knowledge required

Desertification ● ○

Desertification is the spreading of desertlike conditions into regions where vegetation once existed. It generally occurs in arid or semi-arid areas called drylands, where there is minimal rainfall, and is most often the result of poor farming practices (primarily overcultivation), overgrazing, and/or deforestation. These human activities degrade the soil of these marginal regions so that the land can no longer support plant-life.

Drylands constitute almost 40% of the Earth's total land surface. They are ecologically diverse and economically vital in that they serve essentially as the world's breadbasket. Despite their critical role, they are being rapidly destroyed in service of the survival needs of multitudes of impoverished people. According to the United Nations Environment Programme (UNEP), about 75% of the world's drylands are degraded. As can be seen in the accompanying map, desertification is a problem in particular in parts of Africa, Asia, and Central and South America. The problem exists in our Southwest as well.

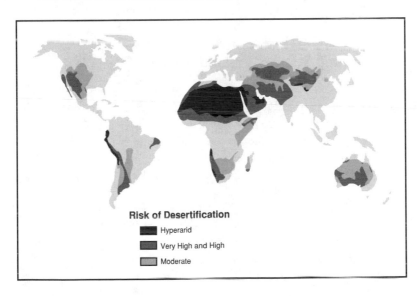

Risk of Desertification

◼ Hyperarid

◼ Very High and High

◻ Moderate

IDEAS TO EXPLORE

What are some of the causes of desertification? Where is it happening most rapidly?

How do drylands contribute to the world's food supply? Which crops are grown?

How is biodiversity being affected? What ecosystems are particularly imperiled? How does land degradation also destroy habitats in water and wetlands? Are cultures being endangered as well?

What roles do drylands play in global climate, energy, and water balance? What are some possible global consequences as a result of degradation of drylands?

What are some solutions that might slow and eventually halt the demise of these globally vital regions?

Many of the world's largest cities, such as Beijing, China, exist in the drylands. What are some of the major problems facing Beijing?

PROJECT

Select and Study a Dryland Region of the World

Identify a dryland region and trace how humans have used the land. Determine how indigenous people survived over the centuries. What foods did they consume? What agricultural practices did they use? How was water conserved? What is the state of the land today? Has desertification or moderate land degradation occurred? What land and water resources are being used? Has biodiversity declined? Have the indigenous people been affected? What are some solutions? For example, the Sahel is a dryland belt existing across Africa in which soil and vegetation have been degraded by virtue of human use of marginal soils.

PROJECT

Ecomigration ● ○

Land degradation has forced people to leave their farms and homes in many regions of Africa, Asia, and Latin America. These ecomigrants often travel to cities where they are then faced with overpopulation and unsanitary living conditions; India and China come immediately to mind as illustrative cases. In Mexico approximately 2,250 square kilometers (875 square miles) of farmland are being degraded annually; this causes thousands of people to flee drylands, often crossing the border to the United States. Identify a region of rapid land degradation, the causes, and the impact on migration trends. What stresses are placed on cities as a result?

RESOURCES

Published Materials

Scientific American, September 1989, p. 78
    P.R. Crosson and N.J. Rosenberg

Scientific American, July 1994, p. 30
    Francesca Bray

National Geographic, Sept. 1984, p. 350
    Boyd Gibbons

Deserts on the March, 4th edition, 1980
    Paul B. Sears
    University of Oklahoma Press, Norman, Okla.

Web Sites

General Information on Desertification
http://www.wcmc.org.uk:80/~dynamic/desert/eng_main.html
Links: Other Interesting Sites on the Internet
http://www.wcmc.org.uk:80/~dynamic/desert/oth_main.html

Bright Edges of the World (an electronic exhibit on drylands from the Smithsonian Institution)
http://www.nasm.edu:1995/asset3.html

Desert Research Institute
http://www.driedu

United Nations Environment Programme (contains the Convention to Combat Desertification)
http://www.unep.ch/

SYMBOL KEY

● Topic of average difficulty

△ Long-term assignment

\# Project requires special facilities, equipment, or supplies

○ Large public or college library required

* Safety precautions required

✚ Highly technical; specialized knowledge required

CHAPTER 3
WATER

Introduction to Water

ABOUT THE EARTH'S WATER RESOURCES

About 71% of the Earth's surface is covered by water. About 97.5% of this water is salt water and is found primarily in oceans. As the pie charts in the figure demonstrate, the fresh water that we so heavily depend on and so rapidly consume represents only 2.5% of the Earth's water, and most of this is unavailable. Most of the fresh water is locked up in the polar ice caps or is too deep inside the Earth. In fact, only about 0.003% of all water is available fresh water in the forms of surface water, groundwater, water vapor, and soil moisture.

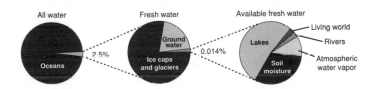

The world's water is found in different bodies and physical states. Most water is found in **oceans; surface water** is found primarily in lakes, streams, and reservoirs; **groundwater** occurs in underground tunnels called aquifers; **water vapor** is mostly in clouds; and some water is in the **solid state** as snow or ice in snowpacks, glaciers, and the polar ice caps. These five locations represent the five parts of the water or **hydrologic cycle— the continuous movement of water between sea, land, and atmosphere.** The hydrologic cycle moves enormous quantities of water between the atmosphere and the Earth's surface. Solar energy causes the water from both the land surface and the oceans to evaporate and rise into the atmosphere where it joins the prevailing winds. The precipitation that then falls onto the land surface flows into nearby streams, rivers, or lakes in a process called **runoff.** Some of the precipitation percolates down into the Earth where it joins an underground aquifer. Eventually most of this water finds its way back to the oceans, the original source of most of the precipitation.

MAJOR ENVIRONMENTAL ISSUES

The hydrologic cycle makes water a renewable resource but a finite one: there is only so much water available at any one time and in any one place. Since the 1950s world water use has grown rapidly, causing water tables to fall, lakes to dry up, rivers to have difficulty making it to the sea, and wetlands to shrink or disappear. Agriculture, industry, and cities are the major water consumers. Currently about 80 countries that collectively represent 40% of the world's population are experiencing water shortages. In addition to the growing demand for fresh water, human activities are polluting water resources. The water cycle described above causes water to come into contact with the land and the atmosphere so that pollutants on the Earth's surface and in the air also end up in our lakes, streams, rivers, and oceans. In addition, we discharge many pollutants directly into our water resources.

As human activities—mostly agriculture, industry, and municipal use—consume and pollute lakes, rivers, streams, wells, and the oceans, aquatic ecosystems are being harmed or destroyed. In fact, aquatic biodiversity is declining at a rate that is four times faster than the decline in terrestrial biodiversity.

About one billion people depend on the aquatic life taken from the oceans and the coastal areas as their main protein source, so it is vital that we begin to view bodies of water not as an unending resource but as part of a life-sustaining system that needs to be protected for our own survival.

Conserving Water ●

PROJECT

Your Personal Water Use Profile ●

It is clear that if our life-support systems on Earth are to remain intact we must conserve energy and water. Even though agriculture and industry are the largest water consumers, reduced household water consumption can play an important part toward conservation, particularly in dry regions like the American Southwest. Recent estimates indicate that each American uses about 80 to 100 gallons daily at home; the greatest household use of water is for toilet flushing, followed by showers and baths. Record for a week your personal consumption of water at home. You can calculate how much running water you use by determining how long it takes to fill a quart container. You can then calculate how much water you consume from faucets. Manuals for appliances often indicate how much water is used for each cycle. Record how much is used for lawns, gardens, and car washes. Also, estimate how much you might save by using less water for toilet flushing and showers or by planting less water-thirsty vegetation in place of your lawn. Ask others to chart their water consumption patterns and then combine results to calculate how much water your community could save in a month or year.

RESOURCES

Published Materials

Overtapped Oasis: Reform or Revolution for Western States, 1989
M. Reissner and S. F. Bates
Island Press, Washington, D.C.

Saving Water in a Desert City, 1989
    William E. Martin et al.
    Resources for the Future, Washington, D.C.

Journal of the American WaterWorks Association (AWWA),
    May 1990, p.22
    Amy Vickers

Garden, July/Aug. 1986, p. 38
    Patricia Wellingham-Jones

Web Sites

U.S. Geological Survey Water Use Homepage
http://h2o.er.usgs.gov/public/watuse/wuqa/home.html

USGS Water Information Homepage
http://h20.usgs.gov/public/watuse/index.html

Waterfront
http://www.mbnet.mb.ca/wpgwater/

SYMBOL KEY

●   Topic of average difficulty

△   Long-term assignment

\#   Project requires special facilities, equipment, or supplies

○   Large public or college library required

*   Safety precautions required

✚   Highly technical; specialized knowledge required

Wetlands ● ○

Although wetlands comprise only about 6% of the Earth's surface, they play a major role in maintaining the stability of the Earth's environment and biodiversity. Wetlands are among the most valuable ecosystems on Earth, performing many vital functions such as trapping sediment and cleansing polluted water, preventing floods, recharging ground water, stabilizing shorelines, and, like rain forests, sustaining a wide variety of plants and animals. In fact, the dense vegetation of wetlands purifies the vast quantities of water that are retained and then slowly released.

In America, wetlands are traditionally grouped into three broad categories: marshes, swamps, and bogs. Prior to the 1970s these vital ecosystems were viewed as wastelands, and millions of acres were drained, dredged, plowed over, and ultimately destroyed. Only about 46% of the wetlands that existed before the United States was founded now remains, with about 300,000 acres disappearing annually. Although today vast areas of wetlands have been set aside and laws have been enacted to protect these precious American resources, there are many citizens who feel that the fate of wetlands that are located on private property should be left to the landowner and that the areas now preserved should be available for development. Wetlands are still threatened today despite our growing understanding of their contribution to our health and well-being.

IDEAS TO EXPLORE

What is the current definition of a wetland? Much controversy has existed over what constitutes a wetland.

What functions do wetlands play in maintaining the global environment? How do they impact on global warming?

What unique characteristics of wetlands sustain biotic diversity? Describe the different kinds of wetlands and their locations.

What are some of the fauna and flora that are supported by these different wetland ecosystems?

PROJECT

Study Wetlands in Your State or Community ● ○

Select a wetland area in your region. How is it classified? What are its characteristics? What kinds of animal and plant life does it support? Perhaps you can isolate just a small site and identify the many different types of species that coexist and describe their interdependence. You can use charts and diagrams to show these relationships. How is that wetland valuable to you? (See some of the ways that wetlands benefit us in the list given in the next project, "Saving Our Wetlands.") You can also show the impact of human activities on the wetland. Are wetlands being lost, polluted, or changed in some serious way?

PROJECT

Saving Our Wetlands ●

Think of ways to educate people about the importance of wetlands to our environment and to our own health and well-being. Use diagrams, graphs, and charts to show why they are important and how they are disappearing. Wetland values include:

- Animals and waterfowl that are used for pelts (alligators, for example) and hunting. The hunting industry earns millions of dollars for many local economies.
- About two-thirds of commercial fish and shellfish species spend some part of their life cycle in wetland habitats, and some important species are permanent residents.

(continues)

- About 82 million acres of wetlands in the United States alone contain marketable timber.
- Wetlands store excess water during periods of heavy rainfall, thereby reducing flooding. Increasing floods along the Mississippi River are associated with the removal of vast tracts of wetland areas.
- Coastal wetlands absorb much of the energy of a storm, thereby protecting shorelines and coastal buildings and other structures.
- Wetlands recharge groundwater. Some major underground rivers or aquifers are becoming depleted because over 50% of Americans use them to supply drinking water and utilize them to obtain water for agriculture and industry.
- Wetlands purify water by acting as sinks for many toxic chemicals that we release.
- Wetlands participate in the global cycles of nitrogen, sulfur, methane, and carbon dioxide and so contribute to the stability of the global environment.

RESOURCES

Published Materials

Wetland Functions and Values: The State of Our Understanding,
Proceedings of the National Symposium on Wetlands,
Lake Buena Vista, Florida
Edited by P. E. Greeson et al.
American Water Resources Association Technical
Publication TPS 79-2, Minneapolis, Minn., 674 pp.

Wetlands, 1986
William J. Mitsch and James G. Gosselink
Van Nostrand Reinhold, New York

Wetlands of North America, 1991
William A. Niering, with photographs by Bates Littlehales
Thomasson Grant, Charlottesville, Va.

Audubon, Nov.-Dec. 1996, p. 42
Ted Williams

Organizations

North American Waterfowl and Wetlands Office
U.S. Fish and Wildlife Service
Arlington Square Building, Room 340
4401 North Fairfax Drive
Arlington, VA 22203
(703-358-1784)

The Nature Conservancy

Web Sites

Lost Wetlands
http://seawifs.gsfc.nasa.gov/ocean_planet.html/peril_wetlands.html

Ocean Planet (an exhibition at the Smithsonian Institution's National Museum of Natural History)
http://seawifs.gsfc.nasa.gov/ocean_planet.html

U.S. Fish and Wildlife Service's National Wetlands Inventory
http://www.nwi.fws.gov/

PROJECT NOTES

Coastal Wetlands ● ○

Coastal wetlands originate along estuaries (inlets of the sea) and beaches and stretch inland. In temperate regions like North America, coastal wetlands include tidal salt marshes, tidal fresh-water marshes, and mangrove swamps (which occur primarily at the southern tip of Florida). There are nine major groups of coastal marshes in the world; three of these are located on the North America–Arctic, Atlantic, and Pacific coasts. In tropical regions of the world, swampy forests of mangroves occur along coastlines.

Coastal wetlands are valuable natural resources. They act as nurseries for many commercial fishes, shrimps, and crabs and serve as nesting sites for waterfowl. Besides their critical service to the fishing industry, coastal wetlands cleanse fresh water by filtering out pollutants; protect inland areas from floods, storms, and associated wave damage; and contribute to the Earth's bio-diversity as do rain forests. Despite their vital role in supporting important food and water resources for humans, and despite their destruction from rising sea levels, coastal wetlands are being drained, dredged, plowed over, and polluted worldwide.

IDEAS TO EXPLORE

Where are coastal wetlands located and how did they form? What are tidal salt marshes? tidal freshwater marshes? mangrove wetlands? How do they differ? What are their unique character-istics?

Describe some of the important animal and plant life that these wetlands support. What characteristics of these important species make them dependent on the existence of coastal wet-lands? Use charts and diagrams to show interrelationships between different plants and animals and their unique environ-ments.

Tidal marshes are particularly fascinating because the life they support is influenced by the alternate flooding and ebbing of the tides. What are some of the mechanisms developed by various life forms that have enabled themselves to adapt to changing water levels?

Why are these ecosystems so valuable to us? Show how all of us benefit from their existence.

Use charts, graphs, and or diagrams to demonstrate the massive destruction of coastal wetlands during the twentieth century. How much has been lost? Where and why? The story of the Everglades in Florida, which is the largest marsh complex in the United States and which once covered more than one million acres, demonstrates the rapid demise of wetlands in North America; since 1920 half of the Everglades National Park has been drained for farms and housing, and much of what remains is now heavily polluted.

RESOURCES

Published Materials

Wetlands, 1986
  William J. Mitsch and James G. Gosselink
  Van Nostrand Reinhold, New York

Salt Marshes and Salt Deserts of the World, 1974
  V. J. Chapman
  J. Kramer, Lehre, Germany

Wet Coastal Ecosystems, 1977
  V. J. Chapman
  Elsevier, Amsterdam

Organizations

North American Waterfowl and Wetlands Office
U.S. Fish and Wildife Service
Arlington Square Building, Room 340
4401 North Fairfax Drive
Arlington, VA 22203
(703-358-1784)

Web Sites

Lost Wetlands
http://seawifs.gsfc.nasa.gov/ocean_planet.html/peril_wetlands.html

Ocean Planet (an exhibition at the Smithsonian Institution's National Museum of Natural History)
http://seawifs.gsfc.nasa.gov/ocean_planet.html

U.S. Fish and Wildlife Service's National Wetlands Inventory
http://www.nwi.fws.gov/

Ocean Voice International
http://www.conveyor.com/oceanvoice.html

SYMBOL KEY	
●	Topic of average difficulty
△	Long-term assignment
#	Project requires special facilities, equipment, or supplies
○	Large public or college library required
✳	Safety precautions required
✚	Highly technical; specialized knowledge required

Marine Pollution ● ○

Ocean pollution is becoming a growing problem that threatens marine biodiversity and contaminates the ocean fish and shellfish that are a major protein source for about one billion people. Many human activities contribute to harmful materials that pollute; in fact, three-quarters of ocean pollution comes from the land and is brought in through the mouths of rivers. An array of industrial chemicals, household cleaning products, gardening chemicals, agricultural chemicals and waste products, and automotive products show up in coastal waters. This polluted runoff from land is called *nonpoint source pollution*. It is interesting to note that the oil spills from the major accidents we hear so much about contribute only about 5% to the total oil pollution of oceans. Debris such as plastics also contribute to ocean pollution. Plastic debris is harmful to marine life that consumes it and to animals that become entangled in it. Raw human sewage entering oceans from sewage-treatment plant overflows, especially during periods of heavy rainfall, threatens human health and closes beaches. Recently, rivers and coastal waters have become overwhelmed by animal wastes from hog and other factory farms when their waste-containment facilities become flooded by heavy rainfall. Mass fish kills have resulted from this pollution, affecting the health and the jobs of those who depend on the rivers and coasts.

IDEAS TO EXPLORE

What are some of the major toxic chemicals that pollute oceans and shorelines? What are their major sources? What is known about their impact on the coastal ecosystems in particular? Present graphs or charts showing the rise in the concentration of these materials over the last several decades.

Why is nonpoint source pollution a growing problem? What can be done to lower its contribution to ocean pollution?

What is the impact of these toxic materials on the health of marine life, including mammals and birds? How are humans affected?

PROJECT

Identify the Watershed You Live in and Determine Your Home's Contribution to Ocean Pollution ●

Everyone lives in a watershed or drainage basin; that is, the precipitation that falls anywhere eventually drains into some body of water such as a stream, river, lake, reservoir, or coastal waters and ultimately ends up in the oceans. Draw a map of your watershed showing the nonpoint and point sources (industries with pipelines, for example) that contribute to water pollution. What toxic materials are drained from your area by your family's activities—use of fertilizers, pesticides, gasoline, oil, household cleaning agents, dog and cat droppings, herbicides, or land clearing, for example? What can you do to decrease your contribution to pollution? Can you decrease your use of toxic materials? conserve water and energy? reduce marine debris? decrease nutrient loads to the coasts? control runoff and erosion?

PROJECT

What Seafoods Do You Eat? ●

According to the National Fisheries Institute, Americans consume billions of pounds of seafood annually, amounting to 15 lbs. per person. Below is a list of the 10 favorite seafoods of Americans and their average per person consumption in 1993:

(continues)

Tuna (canned)	3.5 lb.
Shrimp	2.5 lb.
Alaska Pollock	1.2 lb.
Cod	1.0 lb.
Salmon	1.0 lb.
Catfish	1.0 lb.
Flounder (sole)	0.6 lb.
Clams	0.6 lb.
Crabs	0.4 lb.
Scallops	0.3 lb.

Estimate your or your family's annual intake of seafoods. Where did the seafoods come from? What do these species consume? What contaminants are they exposed to? What is known about these chemicals? Are these fish or shellfish safe to eat?

RESOURCES

Published Materials

Ocean Planet, 1996
   Peter Benchley
   Harry N. Abrams, Inc., and Times Mirror Magazines, Inc., in association with the Smithsonian Institution, Washington, D.C.

The Living Ocean: Understanding and Protecting Marine Biodiversity, 1991
   Boyce Thorne-Miller and John Catena
   Island Press, Washington, D.C.

Organizations

United Nations Food and Agriculture Organization (FAO)

U.S. Fish and Wildlife Service

World Resources Institute (WRI)

Web Sites

Ocean Planet (an exhibition at the Smithsonian Institution's National Museum of Natural History)
http://seawifs.gsfc.nasa.gov/ocean_planet.html

Ocean Awareness
http://www.cs.fsu.edu/projects/sp95ug/group1.7/ocean1.html

Our Living Oceans Annual Report
http://kingfish.ssp.nmfs.gov/olo.html

SYMBOL KEY

● Topic of average difficulty

△ Long-term assignment

\# Project requires special facilities, equipment, or supplies

○ Large public or college library required

* Safety precautions required

✚ Highly technical; specialized knowledge required

CHAPTER 4
THE LIVING WORLD

Introduction to the Living World

Biologists have identified about $1^1/_2$ million different species on the planet. Estimates of the total number of species on Earth range, however, from 10 million to even 100 million. Clearly, we have not even begun to benefit from the remarkable diversity of life forms with which we share our home. Today, as species are disappearing at an unprecedented rate, scientists are scurrying to identify and study new species. The world is right now experiencing its sixth great extinction largely as a result of loss of habitat because of human population expansion and a growing need for food crops; over the past 500 million years there have been five massive extinctions that have wreaked havoc on biodiversity and that have required 20 to 100 million years for recovery. However, this sixth enormous loss of biodiversity, caused by human activities, is occurring at a rate faster than any of the others.

The rain forests of the Southern Hemisphere represent only about 7% of the world's land yet are thought to be home to more than half of all species. Optimistic estimates indicate that species there are disappearing at a rate of 27,000 per year, predominantly because of the clear-cutting of the forests for farmland and ranching; it could be worse. By 1989 less than half of the world's original tropical rain forests remained; studies then showed that they were continuing to be destroyed at a rate of 1.8% annually. Efforts, such as encouraging tourism and planned management for harvesting hardwoods and pharmaceuticals, are underway to convince and assist local communities to preserve the forests.

Many large animals are particularly endangered or are threatened with becoming endangered. The loss of habitat for the African elephant, the cheetah, and the panda, coupled with poaching, has decimated their populations as well as that of many other well-known mammals. Efforts are being made to preserve their numbers by declaring international bans on trade of their body parts as well as by guarding preserves.

Loss of Biodiversity ● ○

Biological diversity or *biodiversity* refers to the rich variety of life forms that thrive on this planet. The three aspects of biodiversity are: the different species that exist in different evironments; the variety of genetic makeup among members of the same species; and the variety of different biological communities like forests, deserts, wetlands, etc. that exist on Earth. In other words, biodiversity refers to the total stock of genes, species, and ecosystems in a region. Although biodiversity is thought by many scientists to be our most precious resource, biodiversity is being severely threatened. Species that have evolved over millions of years are being annihilated at unprecedented rates. Estimates of this destruction are that from 40 to 100 species are becoming extinct *every day*, compared to a natural rate of one species' disappearing every 1,000 years. Whole ecosystems are collapsing and are causing severe economic problems in many areas of the world. Tropical forests that are thought to contain over half of the world's species have been reduced mostly by human activities to less than one-half of their prehistoric cover. It has been estimated that about 20% of the world's species could disappear in just 30 years if the destruction of the rain forests continues at its current rate.

IDEAS TO EXPLORE

What is meant by biodiversity? Does it relate only to the diversity of species?

What is known about the actual number of species on Earth? About how many have we actually studied?

Why is biodiversity so important? Why should pharmaceutical companies be interested in tropical rain forests?

PROJECT

How Each of Us Benefits from Tropical Rain Forests ●

It is difficult for many of us to appreciate the importance of stopping the destruction of the rain forests. After all, we not only live on the other side of the world but our way of life is so drastically different from that of the citizens of nations where the tropical forests are located. The fact is, however, that tropical forests provide habitat for most of the world's animal and plant species from which we take food, wood, fiber, medicines, industrial chemicals, and a variety of raw materials. Rain forests also affect us in less obvious but very important ways. To take just one example, they provide a global sink for carbon; in fact, the destruction of rain forests in the 1980s led to a rapid increase in carbon dioxide (CO_2) in the atmosphere. CO_2 is a greenhouse gas which is thought to increase global temperatures. Because only about 1.4 million species in the world have been studied while there are probably at least 10 million species, most of which reside in the tropical forests, it is clear that we have hardly begun to tap into the products and medicines that occur naturally on Earth. List the products from the tropical forests that you and your family now use. Remember to include raw materials that are used in making the product. Do you think it is helpful to the preservation of the forests if people buy products that come from the rain forests?

PROJECT

The Loss of Biodiversity in Your State ● ○

Research the plants and animals that may have inhabited your state prior to agriculture and buildings. Use maps and graphs to show the different habitats that existed and how they declined. Were there forests, grasslands, or drylands? What animals lived here? What kinds of trees and vegetation were plentiful? How much remains? Have foreign animals and plants been introduced? Did hunting cause some species to disappear? What did your community look like? How can remaining areas be protected and others be restored?

PROJECT

Preserving Biological Hot Spots ● ○

Loss of biodiversity is caused primarily by habitat destruction, introduction of nonindigenous species, and overkill. Habitat destruction contributes most to the decline of biodiversity. "Biological hot spots" are habitats where species that cannot be found anywhere else live and are in danger of extinction. Examples of forested hot spot areas include California's floristic province and San Bruno Mountain, central Chile, the Colombian Choco, western Ecuador, the uplands of western Amazonia, Madagascar, the lower slopes of the Himalayas, Sri Lanka, the Philippines, and southwestern Australia. Other endangered habitats include the Baltic and Aral Seas as well as almost every heavily populated river basin in the world. Coral reefs in the Pacific region, often called the rain forests of the oceans because they harbor such a rich diversity of aquatic life, are being harmed by a variety of human activities.

(continues)

Select a hot-spot region and describe the endemic animal and plant life and ecosystems that are in danger there. With maps and graphs, chart the loss of biodiversity and the causes. What can be done to protect these regions? Ecotourism has been used by Costa Rica to help maintain the 2% of the original rain forest that remains? Is this a possible solution? What are some solutions to stopping loss of biodiversity in the region you have selected?

RESOURCES

Published Materials

The Diversity of Life, 1992
    E. O. Wilson
    W. W. Norton and Co., New York

Scientific American, vol. 261, no. 3, 1989, p. 108
    E. O. Wilson

"Conserving Biological Diversity," in *State of the World,* 1992
    J. C. Ryan
    W. W. Norton and Co., New York

"Reviving Coral Reefs," in *State of the World,* 1993
    Peter Weber
    W. W. Norton and Co., New York

Web Sites

World Resource Institute, Biodiversity Site
http://www.wri.org/wri/biodiv/index.html

Biodiversity Information Network
http://straylight.tamu.edu/bin21/bin21.html

Coral Health and Monitoring Program (CHAMP)
http://coral.aoml.noaa.gov

Environmental Defense Fund (EDF)
http://www.sun-angel.com/edf/edf.html

The Impact of Acid Rain on Plant Growth

Acid rain has damaged most European forests as well as those of the northeastern United States and parts of Canada. It has been estimated that billions of dollars are lost each year because of lower forest productivity caused by sulfur and nitrogen oxides. In addition, acid rain causes significant damage to crop plants.

PROJECT

The Impact of Acid Rain on Tree Seedlings ● △ # ✳

Obtain several tree seedlings in pots from nurseries or transplant into pots seedlings growing in your neighborhood. If you are interested in the effect of acid rain on local forests, use trees that grow naturally in forests: oaks, maples, birch, or evergreens, for example. Take half of your samples and water them with the equivalent of acid rain. Make a solution of sulfuric and nitric acids that yields a pH of 4.0 to 5.0 or less. Rainwater in many areas has been measured with a pH of less than 3.0. Water your control group of plants with water without the addition of the nitric and sulfuric acids. Observe the plants over at least several months, measuring growth of the seedlings in terms of height, length of branches, number of new branches, thickness of trunk and branches, number and health of leaves, etc.

For a more comprehensive project, you can look at several types of trees by watering with your homemade acid rain several seedlings of each type of tree and comparing them with a control group consisting of several seedlings of that species that have been watered with untreated

(continues)

water. You can also extend your project by measuring the impact of different acidities or of different concentrations of sulfuric versus nitric acid. (It may be that one of the acids is more damaging than the other.) Prepare solutions of the same pH but with different relative concentrations of the two acids.

PROJECT

The Impact of Acid Rain on Food Crops ● △ # ✳

A very major concern about the environmental damage associated with acid rain is its impact on the growth of agricultural plants. Select one or more crop plants: tomato, corn, peppers, lettuce, etc. Take several of each and water them with the equivalent of acid rain. Compare these plants with the same plants watered with untreated water. If you perform this experiment during the normal growing season in your region, then you can grow these plants until they produce. Compare them to your controls in terms of rate of growth (height, number of branches, leaves, etc.), resistance to disease and pests, number of flowers, quality of produce, etc.

PROJECT

The Impact of Acid Rain on Soil ● △ # ✳

Recent studies imply that acid rain may have far-reaching consequences on the fertility of soil because it depletes the soil of important nutrients like calcium and magnesium. This is thought to happen because of a simple acid-base reaction between the acids in rainwater and the basic calcium and magnesium compounds that are in the soil.

Once this reaction occurs, the metals become soluble in water and leach out of the soil.

Fill several small, clear plastic containers with soil. Water some for an extended period of time (for example, several months) with the equivalent of acid rain. For the control containers, use untreated water. Plant a fast-growing seed, perhaps bean or corn, on the edge of the container so that you will be able to observe the plant's roots grow. Compare the plants grown in soil watered with acid rain to the controls in terms of days to germination, root structure, plant growth, overall health, etc.

RESOURCES

Published Materials

Ill Winds: Airborne Pollution's Toll on Trees and Crops, 1989
    James Mackenzie and Mohammed El-Ashry
    World Resources Institute, Washington, D.C.

National Acid Precipitation Assessment Program,
    Interim Assessment, vol. 4, 1987
    U.S. Government Printing Office, Washington, D.C.

Web Sites

EPA Office of Air and Radiation
http://www.epa.gov/oar/
Link: http://www.epa.gov/acidrain/ardhome.html
Link: http://www.epa.gov/acidrain/envrben.html

IGC Resources (many links)
http://www.econet.apc.org/acidrain/

Acid Rain
http://www.ns.ec.gc.ca/aeb/ssd/acid/acidfaq.html

Alarming Declines in Ocean Fish ● ○

It has only recently been recognized that the world's oceans and seas are being rapidly depleted of many of their commercial fish as well as of other aquatic life. According to a 1995 report by the United Nations Food and Agriculture Organization (FAO), 9 of the world's 17 major fishing grounds are in serious decline, and 4 have already been depleted of their commercial fish stock. Major industrial nations like the United States and Japan and, in particular, Spain are now arranging to buy fishing rights from cash-poor Third World countries where fish are still plentiful. This exploitation of marine life has stemmed from the unprecedented growth of large-scale fishing vessels equipped with advanced technology; these essentially mine the ocean of its fish and marine life. Advanced navigation aids and miles of automatic nets that detect the approach of fish have turned oceans into factory fish farms, killing not only desirable food fish but also millions of tons of species not wanted for food: dolphins, turtles, and sea birds, for instance. The United Nations estimates that about one-third of the volume of each commercial catch is returned to the ocean, usually dead, because it is unwanted. This is referred to by the industry as "bycatch."

IDEAS TO EXPLORE

Which of the world's major fishing grounds have been "fished out"?

Which commercial fish populations have been decimated and what are the causes? Which aquatic species are being routinely depleted as a result of the bycatch problem of commercial fishing vessels?

How does the collapse of the fishing industry affect the economy of the nations that used the fishing grounds? What are the social consequences?

What is the impact of overfishing on the ocean ecosystem? Can it recover? What needs to be done? How can the bycatch problem be solved? (According to the FOA, bycatch can be quickly reduced by 60%.)

How does the destruction of estuaries and other coastal breeding grounds impact on fish populations?

RESOURCES

See listing following "The Depletion of Fish in U.S. Waters."

SYMBOL KEY

●   Topic of average difficulty

△   Long-term assignment

\#   Project requires special facilities, equipment, or supplies

○   Large public or college library required

✳   Safety precautions required

✚   Highly technical; specialized knowledge required

PROJECT NOTES

The Depletion of Fish in U.S. Waters ● ○

The U.S. National Marine Fisheries Service has indicated that 82% of the commercial stock of U.S. waters has been overfished. The U.S. National Fish and Wildlife Foundation has identified 14 major commercial fish species—representing one-fifth of the world's annual catch—as being so depleted that even with cessation of all fishing it would require at least 20 years for their recovery. In addition, the U.S. fish catch is threatened by the pollution of estuaries in which 70% of fish spend some part of their life cycle.

IDEAS TO EXPLORE

Trace the U.S. fish catch over the last decade. Correlate with dwindling fish populations.

Where in U.S. waters have fish populations suffered most? What has been the economic and social impact? The U.S. Department of Commerce estimates that the collapse of U.S. fisheries is costing about $3 billion annually and 300,000 jobs nationally. What are the causes? What role have pollution and habitat destruction played in the decreasing fish harvest? Can aquaculture (raising fish and shellfish for food) make up the decline in the ocean fish harvest? What can be done to stop the decline in ocean fisheries?

PROJECT

Fishery Collapse ● ○

Select a major fishing ground that has been depleted of commercial fish, the New England ground fishery in the United States, for example. Use graphs and charts to show

(continues)

the rise and fall of the commercial catch over the last two decades as more and more fishing vessels have employed advanced equipment and methods to make them more efficient and more wasteful at the same time. Show the various species that are trapped as bycatch and trace the impact of the industrial fleets on their populations. Suggest solutions to help the region recover. What were the social and economic impacts on the communities that were affected by the collapse of the industry? What political forces permitted the overexploitation of the fishing grounds to begin with?

PROJECT

Bringing Back the Chesapeake Bay ●

The Chesapeake Bay, one of the world's richest estuaries, has been seriously harmed mostly by pollution from nonpoint sources. At one time the bay produced about eight million bushels of oysters annually; now it provides barely one million. Select a specific species of the bay, such as oysters, crabs, or rockfish, and trace its decline and the causes of the decline. Is overfishing a factor? Many groups are working to rejuvenate the bay. How are they curbing the pollution that comes from the heavily populated areas in the regions surrounding the bay? Are there signs of success?

PROJECT

Modern Fishing Techniques and Their Impact on Aquatic Life ● ○

Diagram or find photos of the methods used by modern fishing vessels. These techniques will depend on the fish that is being harvested. One of these is drift-net fishing, in which drifting nets stretch for miles entangling almost all aquatic life that they encounter, harvesting millions of tons of bycatch and reducing ocean biodiversity. Another is purse-seine fishing, which kills large numbers of dolphins. A third is trawler fishing, in which large net trawl bags scrape along the ocean floors, killing unwanted seals and endangered turtles in the process. Determine the nature of advanced techniques that use electronic equipment and satellites to locate and attract large schools of commercial fish. Could you think of ways to alter equipment or procedures so that sea mammals and other unwanted species are spared the fishing nets?

RESOURCES

Published Materials

Ocean Planet, 1996
Peter Benchley
Harry N. Abrams, Inc., and Times Mirror Magazines, Inc., in association with the Smithsonian Institution, Washington, D.C.

The Living Ocean: Understanding and Protecting Marine Biodiversity, 1991
Boyce Thorne-Miller and John Catena
Island Press, Washington, D.C.

Turning the Tide: Saving the Chesapeake Bay, 1991
Tom Horton and William M. Eichbaum
Island Press, Washington, D.C.

Organizations

United Nations Food and Agriculture Organization (FAO)

U.S. Fish and Wildlife Service

World Resources Institute (WRI)

Web Sites

Ocean Planet (an exhibition at the Smithsonian Institution's National Museum of Natural History)
http://seawifs.gsfc.nasa.gov/ocean_planet.html

Ocean Awareness
http://www.cs.fsu.edu/projects/sp95ug/group1.7/ocean1.html

Our Living Oceans Annual Report
http://kingfish.ssp.nmfs.gov/olo.html

SYMBOL KEY

● Topic of average difficulty

△ Long-term assignment

\# Project requires special facilities, equipment, or supplies

○ Large public or college library required

＊ Safety precautions required

✚ Highly technical; specialized knowledge required

Declining Bird Populations ●

In her 1962 book *Silent Spring*, Rachel Carson alerted us to declining bird populations. She attributed the decline to loss of habitat and to environmental pollution. The contributory trends are encroaching at a rapid pace in some regions of the world critical to bird survival. The populations of all migratory songbirds in the mid-Atlantic states decreased by one-half between the 1940s and 1980s. About 250 species of migrating songbirds that breed in the temperate zone of North America and winter in Mexico, Central America, and the Caribbean are declining rapidly. The loss of forests, wetlands, and grasslands are all contributing to their demise. Problems at the beginnings and destinations of their flights are only part of the story; the loss of habitat along the way is also a serious contributing factor. Without suitable habitats on which to land and find food, water, and appropriate shelter, many birds are unable to complete the journey and perish in the attempt.

IDEAS TO EXPLORE

What are the suspected causes of the dramatic decrease in the population of many bird species? Do pollution and alien species play a role in some cases?

Which species are in serious decline? What attempts are being made to help them recover? Have captive breeding programs helped in some cases?

What do you think will be the impact of significant decreases in the bird populations of the Earth? Will insect populations most likely increase, for example?

PROJECT

Design a Backyard Habitat ●

Each of us can turn a backyard, small urban garden, or public park into a suitable habitat for birds and other wildlife. Create designs for your backyard or for any specific outdoor space in your area. Find out what birds are native and what birds pass over along their migration route. Research their food and roosting preferences. Determine what trees, shrubs, and other plants would help them survive. Don't forget to include a water source in your design.

PROJECT

Participate in the Citizen Bird-Monitoring Program ●

The National Audubon Society maintains several ongoing programs that use volunteers to help count and track bird populations. This information serves as an early-warning system for bird species experiencing serious decline. It is hoped that early intervention can prevent these species from becoming endangered. The Christmas Bird Count and the MAPS project (Monitoring Avian Productivity and Survivorship) are two such programs. You can contact the National Audubon Society or your local Audubon organization to find out about these programs. You can also obtain past results and graphically display trends of the bird populations monitored. Which are declining and at what rate? Can you suggest reasons for these trends?

PROJECT

Select an Endangered or Threatened Bird Species ●

Choose a bird like the bald eagle or a bird that is indigenous to your community at some time during the year whose population has been severely reduced. Trace the declining numbers over the last few decades and identify suspected causes of the decline. What can be done to help the species? How many birds are left, and where are they? Have there been recent sightings? Use the Internet to learn of rare bird sightings any place in the world. There are several birding web sites given below.

PROJECT

Track the Migration Path of a Local Bird Species ●

Select a migrating bird from your region and track its migration path. Where does it breed? (For example, does it require forests, wetlands, or grasslands?) Use the World Wide Web to find sightings of your bird species during its migration.

RESOURCES

Published Materials

The Nature Conservancy Magazine,
    vol. 40, no. 1, 1990, p. 4
    D. Wilcove

Silent Spring Revisited, 1987
    Edited by Gini J. Marco et al.
    American Chemical Society, Washington, D.C.

Organizations
National Audubon Society

Web Sites
National Audubon Society
http://www.audubon.org/audubon/contents.html

Cornell Laboratory of Ornithology
http://www.ornith.cornell.edu/Feedback.html

Birdlinks
http://www.phys.rug.nl/mk/people/wpv/birdlink.html

Birding on the Web
http://www.birder.com.

SYMBOL KEY

● Topic of average difficulty

△ Long-term assignment

\# Project requires special facilities, equipment, or supplies

○ Large public or college library required

* Safety precautions required

✛ Highly technical; specialized knowledge required

Using Ethnobotany to Find Useful Drugs ●

About 265,000 flowering plant species are thought to thrive on Earth. Only about 1% of these have been studied with respect to their chemical composition and possible medicinal value.

PROJECT

Use Folklore to Identify a Useful Drug ●

Interview members of your family or community to gain their input about a plant they believe to have medicinal value. Record the details of the folklore and then research the composition of the plant and what is known about its impact on health, if any. You might also seek out members of the same ethnic group as the persons you interviewed to see if the belief leads back to a particular ethnic group or country, or even to a region of a particular country.

If you are unable to uncover family information about plants and healing, you might investigate some aboriginal society like one of the Native American peoples, for example, and research what is known about plants they used for healing.

RESOURCES

Scientific American, vol. 270, no. 6, June 1994, p. 82
  P. Cox and M. Balick

Human Medicinal Agents from Plants, American Chemical
  Society Symposium Series 534, 1993
  Edited by A. Douglas Kinghorn and Manuel F. Balandrin
  American Chemical Society, Washington, D.C.

Roadkill Data

Drivers' trying to avoid hitting animals is the second most common cause of single-car road accidents. The meeting of moving vehicles and wildlife not only causes human injury and property damage but contributes to the decline of species already struggling to survive.

PROJECT

Collect Roadkill Data in Your Community ●

Identify the kinds of animals killed on the roads of your community. Search specified roads each day at consistent times of morning, afternoon, and evening and count the numbers and kinds of animals that have been hit by moving vehicles. Perhaps you can do this for several weeks each season to see if there are changes in the locations, types of animals, and numbers of animals killed. Are some animals killed more at night than at other times? Are there higher roadkills along specific routes and during specific seasons? Use charts, tables, and graphs to show changes in roadkill data with time of day, routes, and seasons. Think of reasons for these differences. Are small animals on their way to water (ponds, creeks, or lakes)? Are they looking for a place to hibernate and seeking forested areas? Could people alter their routes during specific days and times in order to reduce roadkills and accidents?

The Global Decline of the Amphibian Population ● ○

The populations of many amphibian species are declining in numerous regions of the world. Large numbers of deformed amphibians have also been documented in specific wetland areas. The reason(s) for the significant reduction of many species is (are) not known, although it has been suggested that water pollution, acid rain, global warming, and increases in ultraviolet radiation as a result of the thinning ozone layer may all be playing a role. Recently, biologists have documented a serious decline in all seven native toad and frog species that inhabit the relatively pristine Yosemite region; three species have disappeared from the study area entirely. Similar surveys are being conducted at other national parks, including Rocky Mountain, Yellowstone, Grand Teton, Glacier, Mt. Rainier, and Olympic National Park. It is very troublesome that major declines in what were once common species are occurring in relatively pristine environments.

IDEAS TO EXPLORE

Identify some of the species that are in serious decline. Where are they located? What are the suspected causes?

What may be the impact of the loss of amphibians on the ecosystems that they are a part of? Will insect populations, for example, rise? Will the food supply of their predators decrease?

PROJECT

Monitor Local Amphibian Populations ●

Locate a local pond, lake, or stream that amphibians inhabit. Identify the native species and research their

(continues)

behavior and habitat. Select a species and begin a monitoring program. You might concentrate on sites preferred by adults of the species. Knowing the preferred habitat of your species as well as songs and calls will help you find sites where they can be found. Research how your species can best be counted and commence a monitoring program. Contact persons in the field at universities or government or other organizations for advice and to offer your data. Scientists need to gather data from many regions of the world to gain understanding of the causes of this alarming fall in amphibian populations as well as to find solutions to halt it.

PROJECT

Causes of Amphibian Decline ● ○

Research the literature on reports of decreases in the amphibian population in different regions of the world (or select one nation or region). Did researchers find water pollution? Has ultraviolet radiation increased from the thinning ozone layer? Has there been a decrease in the food supply? Is global warming playing a role? Have predator populations increased? Chart the areas of documented decline and the suggested causes. Are there trends that you can see that imply that particular changes are responsible? How can the fall in amphibian numbers be stopped or at least moderated?

RESOURCES

Published Materials

Conservation Biology, Apr. 1996, p. 72
  C. Drost and G. Fellers

Report of the Declining Amphibian Task Force
  Ronald Heyer
  Smithsonian Institution, Washington, D.C.

Organizations

National Biological Service, Fort Collins, Colo.

Web Sites

Froglog
http://acs-info.open.ac.uk/info/newsletters/FROGLOG.html

SYMBOL KEY

- ● Topic of average difficulty
- △ Long-term assignment
- \# Project requires special facilities, equipment, or supplies
- ○ Large public or college library required
- * Safety precautions required
- ✚ Highly technical; specialized knowledge required

CHAPTER 5
ENERGY

Introduction to Energy

Energy production is vital to industrial development and to our own quality of life. For example, we rely on energy for refrigeration and cooking of our food, artificial light for reading, fuel for transportation, and electricity for telephones, televisions, and computers. The list of how we use energy to create a comfortable lifestyle is endless. However, the forms of energy that we use and the ways we use that energy are heavily responsible for the degradation of our environment. The United States is the largest producer and consumer of energy. Our consumption of fossil fuels—oil, coal, and natural gas—is making a significant contribution to the carbon dioxide (CO_2) level in the atmosphere that is in turn causing global warming as well as to the acid rain that is destroying North American forests and numerous lakes and ponds.

Global energy production increased by over 35% in the last 20 years, and most predictions point to continued increases as the world's population rapidly rises and many Third World nations become more industrialized. Most energy in the past came from fossil fuels. With fossil fuel supplies dwindling and the environmental consequences of their continued use looming, many nations will adopt more conservation strategies and will begin to rely more on alternative or renewable energy resources. In the United States, the most wasteful of the industrialized nations today, the application of conservation measures and employment of existing technologies together will generate a major source of energy as well as markedly reducing the harmful environmental impact of fossil fuels. Shifting to alternative energies that use renewable resources, like biomass, solar, wave, wind, geothermal, and small-scale hydro, will significantly reduce our impact on the life systems that support us. In fact, in many regions of the world alternative energy resources are replacing fossil fuels.

Third World nations in particular are turning to renewable resources. With their exploding populations and desire to industrialize, these countries are looking to rely on cheaper and more available local sources of energy. Without the extensive reserves of fossil fuels typical of so many industrialized nations, many developing countries are trying to build self-reliant economies for which access to energy resources is a necessity.

SYMBOL KEY

● Topic of average difficulty

△ Long-term assignment

\# Project requires special facilities, equipment, or supplies

○ Large public or college library required

* Safety precautions required

✚ Highly technical; specialized knowledge required

Energy Conservation ●

The United States, with only about 5% of the world's population, uses 30% of the world's energy resources and produces 25% of all of the global CO_2 emissions. On the average, each of us generates 44,000 lb. of CO_2 annually. We are making an enormous contribution to global warming and acid rain as a result of our prodigious use of oil, coal, and natural gas. We can, however, markedly decrease our consumption of fossil fuels, save money, create jobs, and lower air pollution by conserving energy. It has been estimated that about 40% of the energy used in the United States is wasted, primarily by inefficient vehicles and leaky buildings. For example, if we were to use ceiling fans and double-glazed windows, we would save more oil than what is estimated optimistically to be in the entire Arctic National Wildlife Refuge. In fact, if we had been as energy efficient as Japan or Sweden in the last decade, we would have been spending $200 billion a year less in energy costs, an amount about equal to the federal deficit. By adopting the energy conservation measures and technologies used by Japan and many Western European nations, we could significantly reduce our reliance on fossil fuels and our substantial contribution to air pollution. At the same time, we could improve our economy and create more jobs as research and development grow.

IDEAS TO EXPLORE

What are the main ways that the United States can conserve energy? How can improved systems of mass transit help? Are large gas-guzzling cars a significant part of the problem?

How can buildings become more energy efficient? Features like heavy masonry construction, night flushing, daylighting and light controls, and shading from plants and trees have reduced energy consumption in efficient buildings. What can be done to buildings already erected to decrease their energy costs?

All of us must learn to lessen our impact on the environment. What lifestyle changes can we adopt to reduce our consumption of energy? How can you make your home more energy efficient?

Since 1978 many nations have decreased their energy consumption by conservation. Why have we not followed their lead? What government policies would encourage conservation?

PROJECT

Your Personal Energy Profile ●

For at least one week, record each time you consume energy: each time you turn on lights, heat or cool your home, drive in a car, watch television, use home appliances, etc. If you know the power output of the appliance being used, and if you record the amounts of time involved, you can estimate your total personal energy consumption. Analyze your energy consumption patterns and think of how you can make changes in your behavior to reduce your personal energy usage. Can you use fewer lights or more efficient ones? Is the television or radio on without anyone really paying attention to it? Can you use your washing machine less by washing larger loads? Can you use less hot water? Is mass transit or car pooling a possibility?

PROJECT

Energy Analysis of Your Home ●

Heating, cooling, and lighting of buildings amounts to about one-third of all energy consumption in modern nations. A significant amount of this consumption is attributable to waste. For example, one-third of heated air in American homes and buildings escapes through windows, cracks, and holes. It has been estimated that this costs consumers $13 billion annually. Moreover, this wasted energy is equivalent to all of the energy contained in the oil flowing through the Alaskan pipeline each year.

Determine how your home can be more energy efficient. Does your home need weatherization measures like insulation or new energy-efficient windows to decrease heating and cooling costs? Can you design a landscape that will reduce energy consumption? Are your lights energy efficient? Are your home appliances the most energy efficient? Many electric power companies across the nation are heavily involved in helping their customers to conserve. Contact your local power company for information relevant to this project.

PROJECT

Energy-Efficient Homes and Buildings ● ○

Design an energy-efficient home for your community. Make use of local renewable energy sources like wind or solar energy. Find out about the many efficient designs that are being used today. Homes are now being built in Canada that consume one-tenth the energy of an American home. The Georgia Power Company office building in Atlanta uses 60% less energy than most conventional buildings of its size; its largest surface area faces south to capture solar energy. Each floor shades the one below to block out the high summer sun yet lets in the lower-lying winter sun. Energy-efficient lights are focused on individual work stations rather than entire rooms.

Michael Reynolds, a builder in Taos, New Mexico, uses sunlight for heating and lighting and the sides of hills for insulation. Michael Sykes patented a solar envelope house that is heated and cooled passively by solar energy. It is sold in kit form and can be used in most regions of the United States. Solar designs, in general, can provide about 70% of residential heating requirements. Currently, there are over 250,000 passive solar homes in North America. Straw-bale houses, popular in the 1800s in Nebraska, are making a comeback in Canada, Finland, and Mexico and in many area in the United States. They rely on thick walls constructed from post-and-beam structures with stacked bales of straw covered with plaster or adobe.

(continues)

You can begin by consulting your local electrical power company. Many of these companies across the nation are helping people design energy-efficient homes.

PROJECT

Using Green Lights in your School ● ○

Standard incandescent light bulbs are very energy wasteful; they are only 5% efficient and last only 750 to 1,500 hours. Efficient fluorescent light bulbs and electronic lamp bulbs use one-quarter as much electricity and last 10 to 20 times longer. Calculate how much your school would save if it replaced inefficient bulbs with these modern energy savers. The U.S. Environmental Protection Agency (EPA) has, in fact, started a Green Lights program to encourage energy-efficient lighting.

RESOURCES

Published Materials

A Consumer Guide to Home Energy Savings
American Council for an Energy Efficient
Economy (ACEEE)
1001 Connecticut Avenue NW, Suite 801
Washington, DC 20036

Solar Houses: 48 Energy-Saving Designs, 1978
L. Group
Pantheon, New York

Earthship: How to Build Your Own, 1990
Michael Reynolds
Solar Survival Press, Taos, N.M.

The Straw-Bale House, 1994
Athene Swentzell Steen et al.
Chelsea Green Publishing, White River Junction, Vt.

Living in the Environment, 9th edition, 1996
G. Tyler Miller, Jr.
Wadsworth Publishing Company, New York

The Ecology of Commerce, 1993
Paul Hawken
HarperBusiness, HarperCollins Publishers, New York

"Designing a Sustainable Energy System,"
in *State of the World,*
1991
C. Flavin and N. Lenssen
W. W. Norton and Co., New York

Energy Policy, vol. 20, no. 6, 1992, p. 547
Gregory Kats

Energy Policy, vol. 19, no. 10, 1991, p. 953
A. Reddy

Organizations

U.S. Department of Energy (DOE)

U.S. Environmental Protection Agency (EPA)

International Institute for Energy Conservation
Washington, D.C.

American Council for an Energy Efficient Economy (ACEEE)
1001 Connecticut Avenue NW, Suite 801
Washington, DC 20036

Web Sites

The Energy Outlet
http://energyoutlet.com/

Energy Links
http://www.epri.com/energylinks.html

Energy Program Library
http://www.energy.wsu.edu/ep/library/

PROJECT NOTES

Renewable Energy Resources ● ○

The world's demand for energy is expected to increase significantly in the future, especially in developing nations. As these
countries seek to raise their standards of living and expand their
economies, they will require extensive energy services. According to the U.S. Office of Technology Assessment, commercial
energy consumption may triple over the next 30 years, increasing by 14% the contribution of developing nations to global
commercial energy use. Demand for electricity by nations of the
Southern Hemisphere are predicted to rise by 7% to 10% per
year over the next few years. If the energy demands of both
developed and undeveloped nations continue to be met primarily by nonrenewable fossil fuels—oil, coal, and natural gas—the
environmental consequences may be catastrophic; global warming, acid rain, air pollution, and water pollution will threaten
our life-support systems on this planet. In addition, the global
supply of oil, the world's most widely used resource, will most
likely run out in the next 40 to 80 years. It is clear that the world
must shift to a new mix of renewable energy resources. These
alternative energies include biomass, wave, wind, geothermal,
solar power, and small-scale hydro. Use of the last four of
these—wind, geothermal, solar, and small-scale hydro—is
spreading and shows much promise for the future.

The United States, Denmark, and Australia are major pro-
ducers of wind power.

IDEAS TO EXPLORE

By how much is global energy demand predicted to increase? In
what regions of the world will demand increase the most?

Will fossil fuels continue to be the main source of this predicted
energy consumption? What environmental problems are caused
by the use of fossil fuels for energy production?

What are some renewable energy sources? Are some of these alternative energies being used today? Select one or two alternative energies and discuss how they are being used today, along with plans for future expansion. What are the advantages and disadvantages of using these alternative energies?

RESOURCES

Published Materials

"Designing a Sustainable Energy System,"
   in *State of the World,*
   1991
   C. Flavin and N. Lenssen
   W. W. Norton and Co., New York

Renewable Energy: Today's Contribution, Tomorrow's Promise,
   1988
   Worldwatch Paper 81
   C. P. Shea
   Worldwatch Institute, Washington, D.C.

Renewable Energy: Sources for Fuels and Electricity, 1992
   T. B. Johansson et al.
   Island Press, Washington, D.C.

Web Sites

California Energy Commission
http://www.energy.ca.gov/energy

Center for Renewable Energy and Sustainable Technology (CREST)
http://crest.org/

Department of Energy, Energy Efficiency and Renewable Energy Network
http://www.eren.doe.gov/

CHAPTER 6
HUMAN ISSUES

Introduction to Human Issues

For centuries humankind has viewed the environment as an unlimited source of fuels, minerals, water, soil, and animal and plant life. We have developed very sophisticated methods of mining these resources without regard to the impact of our use upon the life-support systems that we need to live. The consequent pollution of the air, water, and land has not only harmed human health and caused many thousands of deaths yearly in the United States alone, but also is decimating trees, crop plants, fish, and wildlife in many regions of the world. It is clear that in order to develop a sustainable future and protect our health and long-term survival, we must protect the air, water, land, and living world of which we are just a small part.

SYMBOL KEY

● Topic of average difficulty

△ Long-term assignment

\# Project requires special facilities, equipment, or supplies

○ Large public or college library required

* Safety precautions required

✚ Highly technical; specialized knowledge required

Environmental Justice ● ○

The term *environmental justice* refers to the premise that all people regardless of age, culture, race, gender, or socioeconomic class should be adequately protected from environmental hazards. However, the reality is that municipal landfills, incinerators, toxic-waste dumps, and polluting factories and industries tend to be located in neighborhoods in which largely people of color (African Americans, Latinos, Native Americans, Asians, and others) and working-class people reside. Members of these communities are generally exposed to more polluted air and/or polluted water as a result of their proximity to one or more of these polluting structures. Members of racial minority groups and working-class people are more likely to suffer from unsafe working conditions as well. For example, 300,000 farm workers in the United States, most of them minority group members, are exposed to high levels of pesticides; and Navajo Indians inhale dangerous levels of radioactive dust in the uranium mines of Arizona along with drinking water containing radioactive metals and gases that come from the mines. Neighborhoods of mainly African Americans located in the Lower Mississippi Delta region have such high rates of cancers that the region is called "Cancer Alley." Mercury, pesticides from agricultural runoff, harmful chemicals from Superfund sites, and chemical spills are just some of the pollutants that degrade both the air and the water of these neighborhoods.

IDEAS TO EXPLORE

What is meant by environmental equity, environmental racism, or environmental discrimination? What are some of the problems in proving its existence?

What health disparities between ethnic and racial minorities and whites are thought to be related to greater exposure to environmental health hazards?

What political, economic, and social factors cause the poor and minority populations to bear a disproportionate share of environmental health hazards?

What is being done to promote environmental equity? (For example, on February 11, 1994, President Clinton signed Executive Order 12898 directing appropriate federal agencies to develop policies and programs that will prevent any community or specific population from bearing an inordinate share of the adverse health and other effects of pollution. What have they done so far?)

Can you suggest some intervention strategies that can improve the health status of affected groups?

PROJECT

Investigate Examples of Environmental Racism in Your State, City, or Community ●

Locate the sites of landfills, toxic waste dumps, incinerators, or polluting factories or industries in your state, city, or community. To learn about toxic waste sites, inquire about Superfund sites from your state or federal EPA (Environmental Protection Agency). Draw a map of the area surrounding these sites and research the socioeconomic status and the ethnicities/races of the residents of these neighborhoods. What are the environmental problems associated with the site or structure? How long has it been there? What procedures were followed in approving its presence? What are the governing zoning regulations and how do they compare with those of other communities in the state? How does the site or structure affect the economic well-being of the surrounding communities? What political, social, and economic factors led to its presence in that particular community as opposed to in another community?

For a more extensive project, also research lead in the bloodstream of children.

PROJECT

Lead in the Bloodstream of Children ●

There is much evidence to support the notion that children of African-American and Mexican-American families are more likely to have higher blood lead levels than do white children. Determine if this trend applies to your state or community. Locate hospitals or medical offices that serve neighborhoods that are home to people of different race and socioeconomic status and see if they will provide you with results of lead tests. Contact the Centers for Disease Control (CDC) and request appropriate information for your project.

RESOURCES

Published Materials

Health and Environment Digest (a publication of the Freshwater
   Foundation), vol. 9, no. 9, 1996, p. 73
   K. Sexton

Toxicology and Environmental Health,
   vol. 9, no. 5, 1993, p. 843
   K. Sexton et al.

Environmental Science and Technology,
   vol. 22, no. 1, 1995, p. 69
   Perlin et al.

Get the Lead Out
   Catherine McVay Hughes and Chris Meyer
   New York Public Interest Group Publications, 1995
   9 Murray St., 3rd Floor
   New York, NY 10007-2272

Organizations

U.S. Census Bureau

Centers for Disease Control (CDC)

Web Sites

Environmental Protection Agency
http://www.epa.gov/

National Institute of Environmental Health Sciences (NIEHS)
http://www.niehs.gov/

Centers for Disease Control
http://www.cdc.gov

SYMBOL KEY

● Topic of average difficulty

△ Long-term assignment

\# Project requires special facilities, equipment, or supplies

○ Large public or college library required

* Safety precautions required

✚ Highly technical; specialized knowledge required

Incorporating Environmental Issues in Liberal Arts Curricula

The health of our environment is an issue that is creeping into many aspects of our lives. We are becoming more and more aware of the importance of protecting our water, land, soil, and biodiversity as we realize that our health and well-being are inextricably linked to the health of the environment. To help us make informed decisions about how we interact with our environment, many academicians are including environmental issues in their courses.

PROJECT

Green Curricula ●

Select a course or area of study and design a curriculum that incorporates environmental issues.

RESOURCES

Chronicle of Higher Education, Feb. 23, 1996, p. B1
  J. Collett and S. Karakashian

PROJECT NOTES

Effects of Ultraviolet-B (UV-B) Radiation on Health ● ○

Numerous studies have shown that the ozone layer in the stratosphere is being depleted, primarily as a result of the release of chloroflourocarbon compounds (CFCs) into the atmosphere. Although large holes in the ozone layer over the poles have been detected for some time, only recently has ozone depletion been measured over the populated areas of the Northern and Southern Hemispheres; the largest decreases appear to be in the Southern Hemisphere. The concentration of ozone gas in the stratosphere is a factor in determining the amount of UV-B radiation (UV light with wavelengths between 280 and 320 nm) reaching the Earth's surface. A 1% decrease in stratospheric ozone is estimated to cause a 1.25% to 1.5% increase in UV-B radiation at ground level, while a greater than 5% depletion is estimated to exponentially increase the high-energy UV-B rays reaching the Earth. Given the long lifetime of CFCs in the atmosphere, the ozone layer is expected to continue to thin and, even with significant reductions in current CFC emissions, we will be exposed to ever-greater amounts of UV-B. What impact will these energy light rays have on human health?

Evidence from different types of studies—clinical observation, epidemiological, and experimental—indicate that greater exposure to UV-B radiation can have a serious effect on human health; risks for skin cancer are increased with greater exposure as is a suppression of the immune system that can have far-reaching consequences. Two kinds of skin cancers have been associated with UV exposure: nonmelanoma skin cancer (NMSC) and cutaneous malignant melanoma (CMM). Both of these skin cancers occur more in southern latitudes: the highest rates for the Northern Hemisphere are in the southern United States; Australia also has very high rates of skin cancer. A study in Australia showed that protection of children from exposure to UV rays decreased their rate of skin cancer in adulthood. The recent rise in rates of skin cancers is being linked to ozone depletion; for example, a 1% loss in stratospheric ozone has been estimated to increase the incidence of nonmelanoma skin cancer by 2%.

Ultraviolet light has also been associated with suppression of the human immune system, decreasing the body's ability to fight infection. As UV levels rise, concern grows over already immunocompromised patients and questions are being raised about the effectiveness of vaccinations used to prevent crippling diseases in children. Will some children not be able to appropriately respond to the vaccinations because of a suppression of their immune systems by increased exposure to UV-B radiation?

IDEAS TO EXPLORE

What evidence is there to support the concern that ozone layer loss will lead to higher rates of skin cancers? Where are these expected to occur in particular? What are some other factors that play a role in the quantity of UV-B radiation that reaches ground level?

What precautions should be taken, especially with children, to reduce risk of skin cancers?

What data imply the weakening of the human immune system with exposure to UV light? Will there be a rise in infectious diseases as a result of ozone layer depletion?

RESOURCES

Published Materials

Health and Environment Digest (a publication of the Freshwater Foundation), vol. 9, no. 3, July 1995, p. 22
R. Gray and K. Cooper

Photochemistry and Photobiology, vol. 50, no. 4, 1989, p. 507
F. Urbach

United Nations Environment Programme (UNEP), 1991, p. 15
J. D. Longstreth et al.

Proceedings of the National Academy of Sciences USA, vol. 89, 1992, p. 8497
K. D. Cooper et al.

Web Sites

Stratospheric Ozone and Human Health
http://sedac.ciesin.org/ozone/

SYMBOL KEY

● Topic of average difficulty

△ Long-term assignment

\# Project requires special facilities, equipment, or supplies

○ Large public or college library required

* Safety precautions required

✚ Highly technical; specialized knowledge required

Diet and Cancer ● ○

There is a growing body of evidence that suggests that a large proportion of human cancers are influenced by dietary factors. In fact, some major studies point to diet as the causative factor in 35% of all cancers while attributing tobacco as the culprit in 30% of all cancers. Modern diets, in affluent societies in particular, consist of a multitude of natural and synthetic chemicals many of which are suspected carcinogens and others of which are thought to be anticarcinogens. There is considerable evidence, for example, that a diet rich in fruits and vegetables is associated with a decreased risk of many cancers and of heart disease. Vegetarians live longer and enjoy better health. There has been much research that implicates saturated fats as dietary factors that increase the risk of breast, colon, and prostate cancer. The myriad of synthetic chemicals in our diet is cause for concern. Many of these substances, particularly the halogenated organic chemicals, are components of herbicides and pesticides that even in small quantities are known to be carcinogenic to rodents. Because many of these organic chemicals—like DDT, dioxins, and PCBs (polychlorinated biphenyls)—bioaccumulate in our fat tissue, they are also present in human milk and can thus be passed on to nursing infants. The impact of the combination of these chemicals on cancer risk is unknown.

IDEAS TO EXPLORE

Research the recent literature on the relationship between fruits and vegetables and cancer risk. Recently the U.S. Department of Agriculture began supporting programs to put more fruits, vegetables, and grains into school lunches. What do we know about the relationship between different food habits of people around the world and cancers? What is known about the Seventh Day Adventists and the Mormons in terms of low cancer rates?

What are some of the potentially hazardous herbicides and pesticides in our diets? What is known about them? Have they all been adequately tested? How does the Delaney clause relate to food safety? A recent study by the National Research Council on pesticide residues in the diets of infants and children indicated that infants and children constitute a population that is not being adequately protected. What action has the federal government recently taken to help ensure food safety? What steps do you think should be taken to decrease our exposure to potentially harmful chemicals in the diet? Make a list of the synthetic chemicals present in our diets, their sources, and what is known about their impact on human health.

PROJECT

Pesticides in Your Diet ●

Record everything you consume for one week. What herbicides, pesticides, or fungicides were used in producing each food item? For a meat product, think about the animal feed used? Was it sprayed with pesticides that might accumulate in the animal's tissues?

RESOURCES

Published Materials

Health and Environment Digest (a publication of the Freshwater
   Foundation), vol. 10, no. 5, Sept. 1996, p. 33
   R. Estabrook

Carcinogens and Anticarcinogens in the Human Diet, 1996
   National Research Council
   National Academy Press, Washington, D.C.

Science,
   vol. 272, 1996, p. 1489
   S. F. Arnold et al.

Pesticides in the Diets of Infants and Children, 1993
    National Research Council
    National Academy Press, Washington, D.C.

The Safe Shopper's Bible, 1995
    David Steinman and Samuel S. Epstein
    Macmillan, New York

Organizations

Centers for Disease Control (CDC)

Web Sites

National Association of Physicians for the Environment (NAPEnet)
http://intr.net/napenet/

National Institute of Environmental Health Sciences (NIEHS)
http://www.niehs.nih.gov/

SYMBOL KEY

● Topic of average difficulty

△ Long-term assignment

\# Project requires special facilities, equipment, or supplies

○ Large public or college library required

* Safety precautions required

✚ Highly technical; specialized knowledge required

Global Warming and Infectious Disease

According to a recent report by the United Nations Intergovernmental Panel on Climate Change, the redistribution and resurgence of infectious disease, unprecedented since the 1880s, may be due to the impact of global warming.

PROJECT

Impact of Global Warming on Infectious Disease
● ○

Select an infectious disease like cholera, malaria, or yellow fever and estimate the impact of global warming on its prevalence. Will the disease spread to regions where it was formerly infrequent, perhaps as temperate zones warm up? Will it occur more frequently in tropical regions? How does temperature impact the life cycle of disease carriers?

Select an infectious disease that is prevalent in your community, the flu or strep or another bacterial infection. How will climate changes affect the distribution and incidence of the disease? Other diseases of concern include hantavirus (which spread in the Southwest in 1993 after an unusually wet period) and lyme disease (which is spreading from the Northeast where it is most prevalent into the South and West). Along with an increase in the average annual global temperature, many regions are expected to experience harsher winters and hotter summers. How will these changes affect the spread of disease?

RESOURCES

Earth, Apr. 1996, p. 20
  T. Yulsman

American Journal of Public Health, Feb. 1995, p. 168
  P. Epstein

SYMBOL KEY

● Topic of average difficulty

△ Long-term assignment

Project requires special facilities, equipment, or supplies

○ Large public or college library required

* Safety precautions required

+ Highly technical; specialized knowledge required

Particulate Air Pollution and Public Health ●

Recent studies indicate that airborne particulates pose a very serious danger to public health. Although air pollution has long been linked with respiratory disease, recent research now shows a strong association specifically between concentrations of particulates and deaths from cardiovascular disease, chronic obstructive lung disease, asthma, and pneumonia. In particular, a strong link has recently been shown between fine particulate concentrations (10 micrometers or less) in air and daily death rates. The landmark Harvard Six City Study related long-term average particulate concentrations in city air to life expectancy; the strongest association was with fine particulate concentrations. In addition, toxicological evidence adds to the extensive and growing evidence that particulate air pollution is a major health hazard: significantly elevated mortality rates were seen in bronchitic rats exposed to particulate air pollution for six hours for each of three days, while little change was observed for the control rats. In this study inflammatory mediators were isolated in both the lungs and the hearts of the bronchitic rats.

The Natural Resource Defense Council estimates that 56,000 lives in 239 cities are lost annually because of small-particle pollution; more than 5,000 of these are in Los Angeles and 4,000 in New York City. The Environmental Protection Agency and the Harvard School of Public Health suggest that between 50,000 and 60,000 deaths are caused yearly by airborne fine particulates. Consequently, they strongly urge the introduction of a federal rule that will limit the permitted concentration of fine particulates in outdoor air. Fine particulates are generated from several sources: coal-fired power plants and industrial burners are the major polluters; gasoline- and diesel-burning cars and trucks, wood-burning stoves, incinerators, and other fuel-burning devices add to the pollution; and agricultural activities like large feedlots and livestock waste also contribute.

Aside from enforcing stricter Ambient Air Quality Standards, much can be done to reduce fine-particle pollution. For example, large-scale reductions in fuel and electricity usage could significantly decrease the concentrations of these deadly contaminants in the air. If we all insulate our homes properly and use energy-efficient lighting and appliances as well as more energy-efficient cars, we can mitigate air pollution.

IDEAS TO EXPLORE

What evidence is there that fine-particle air pollution is a major hazard to public health?

What causes this type of air pollution?

Who is most affected by airborne particulates?

What are some solutions? Should the federal government regulate particulate concentrations in outdoor air? What can electric utilities and other industries do to reduce their emissions of this harmful pollutant?

PROJECT

Airborne Particulates in Your Community ●

Monitor the fine-particle pollution either in your community or in some other selected area by contacting local pollution-control authorities. Does fine-particle pollution change with the weather conditions and the time of day? Does it vary with the seasons? Use graphs and charts to show trends. Where are the particles coming from? Compare your averages with those in other areas. (The Environmental Protection Agency and the Natural Resources Defense Council have data on the average fine-particle concentrations for numerous cities.)

RESOURCES

Published Materials

New England Journal of Medicine, Dec. 9, 1993, p. 1753
    D. W. Dockery et al.

"More Than 50,000 Die Each Year From Air Pollution,
    Environmental Group Reports,"
    The New York Times, May 9, 1996
    Philip J. Hilts

Breath Taking, 1996
    A report by the Natural Resources Defense Council
    (NRDC)

Web Sites

Office of Air and Radiation, Office of Mobile Sources
http://www.epa.gov/OMSWWW/
Links: http://www.epa.gov/OMSWWW/o5-autos.html

Air Radiation and Toxics Division (ARTD)
http://www.epa.gov/reg3artd
Links: Clean Cities
http://www.epa.gov/reg3artd/partner/clncit.html
Links: Protecting Our Air Quality
http://www.epa.gov/reg3artd/airqual/hmpg.html

National Association of Physicians for the Environment (NAPEnet)
http://intr.net/napenet/

National Institute of Environmental Health Sciences (NIEHS)
http://www.niehs.nih.gov/

Xenoestrogens in the Environment and Brest Cancer Risk ● ○

Numerous studies link exposure to carcinogenic chemicals in the environment with various cancers. Although about 44,000 American women die each year from breast cancer, more than succumb to AIDS, only recently have researchers turned their attention to the potential role of certain chemicals in the environment that behave like estrogen in the body—the so-called xenoestrogens. Estrogen levels in the body have been linked with breast cancer, so xenoestrogens are of particular concern. Xenoestrogens may be increasing breast cancer risk by effectively increasing internal exposure to estrogen. Some environmental xenoestrogens include such pesticides as toxaphene, dieldrin, DDT, methoxychlor, and endosulfan and industrial chemicals such as benzene, polychlorinated byphenols (PCBs), polyaromatic hydrocarbons (PAHs), atrazine (herbicide), nonylphenol, and related phenolics.

During the past three years, several major studies have been conducted to assess the role of estrogenic chemicals in the development of breast cancer. In one study PCBs and DDT metabolites were found in higher levels in the mammary adipose tissue of breast cancer patients than in patients with benign disease. In another study Dr. Mary Wolff and research associates reported significantly elevated levels of DDE (metabolite of DDT) in the serum of breast cancer victims than in a disease-free control group. Xenoestrogens have also been shown to cause mammary tumors in laboratory animals.

IDEAS TO EXPLORE

What recent studies suggest a link between risk of breast cancer and exposure to xenoestrogens?

How do these chemicals enter the body? Is diet the main pathway for most people? Do the diets of meat-eaters contain larger amounts of xenoestrogens than those of vegetarians?

Do breast cancer rates differ from country to country? What happens when women move from nations of low cancer rates to the United States where the rate of breast cancer is high? Is your occupation important in assessing risk of the disease?

PROJECT

Assess Your Exposure to Xenoestrogens ●

Determine which xenoestrogens you and others in your community are exposed to daily or weekly. Do adults consume more or less of these than do children or teenagers? How about older adults? You can also compare exposure of men and women. Is there exposure mainly from the diet or are these chemicals in the air and water supply? An FDA survey, for example, indicated that the average 16-year-old American male consumes about 0.8 micrograms of the pesticides endosulfan, dieldrin, and methoxychlor and 1.7 micrograms of DDT metabolites each day. There is presently much concern over the impact of pesticides on human health; in August 1996 President Clinton signed the Food Quality Protection Act to address some of these issues. The law requires the U.S. Environmental Protection Agency to publish and distribute annually a brochure discussing pesticides, their benefits and risks, and how to reduce exposure to them.

PROJECT

"Hot Spots" in Your State ●

Research breast cancer rates in the regions of your state. Contact the Centers for Disease Control, the National Cancer Institute, a state health agency, or perhaps a university that is studying breast cancer risks. Are there regions where rates are higher than others? How do rates of breast cancer compare with those in the rest of the nation? Are there industrial activities in the high-rate regions? What might be other risk factors that are contributing to the high occurrence of the disease? Cornell University in Ithaca, New York, for example, is assessing risk factors for breast cancer for the women of New York State because of the large number of clusters or "hot spots" where the disease rates are very high. Both consumers and researchers can access the information primarily through a World Wide Web page. To find out more about this first attempt to gather and assess all of the studies on the environmental risk factors of breast cancer, contact Roger Segelken, Cornell University News Service, at 607-255-9736.

RESOURCES

Published Materials

Health and Environment (a publication of the Freshwater Foundation), vol. 8, no. 10, 1995, p. 77
J. Melius and S. Safe

Science, vol. 259, 1993, p. 616
E. Marshall

Archives of Environmental Health, vol. 47, 1992, p. 143
F. Falck et al.

National Cancer Institute, vol. 85, 1993, p. 648
    M. S. Wolff

Medical Hypothesis: Environmental Health Perspective,
    vol. 101, 1993, p. 372
    D. Davis et al.

The Safe Shopper's Bible, 1995
    David Steinman and Samuel S. Epstein
    Macmillan, New York

Web Sites

OncoLink: University of Pennsylvania Cancer Resource
http://cancer.med.upenn.edu/

SYMBOL KEY

● Topic of average difficulty

△ Long-term assignment

\# Project requires special facilities, equipment, or supplies

○ Large public or college library required

* Safety precautions required

✚ Highly technical; specialized knowledge required

PART III

RESOURCES

Resources for Research

Below are listed scientific societies, regional, national, and international organizations, and government agencies that may be helpful in researching your project or report. Also, a list of environmental databases, general web sites on the environment and scientific supply companies has been provided. You can use these web sites to begin researching most topics on the environment.

SCIENTIFIC SOCIETIES

AMERICAN ASSOCIATION FOR THE ADVANCEMENT OF SCIENCE

AMERICAN CHEMICAL SOCIETY

AMERICAN GEOPHYSICAL UNION

AMERICAN INSTITUTE OF BIOLOGICAL SCIENCES

AMERICAN LITTORAL SOCIETY

AMERICAN NATURE STUDY SOCIETY

AMERICAN PHYSICAL SOCIETY

AMERICAN PUBLIC HEALTH ASSOCIATION

AMERICAN SOCIETY FOR AGRONOMY

AMERICAN SOCIETY FOR MICROBIOLOGY

AMERICAN SOCIETY OF NATURALISTS

AMERICAN SOCIETY OF PLANT PHYSIOLOGISTS

AMERICAN SOCIETY OF ZOOLOGISTS

AMERICAN VETERINARY MEDICAL ASSOCIATION

BIOPHYSICAL SOCIETY

GEOLOGICAL SOCIETY OF AMERICA

SOCIETY FOR EPIDEMIOLOGIC RESEARCH

SOCIETY FOR RESEARCH IN CHILD
DEVELOPMENT

SOCIETY OF TOXICOLOGY

SOCIETY FOR VECTOR ECOLOGY

The addresses of the professional societies listed above can be
obtained from the:
AMERICAN ASSOCIATION FOR THE ADVANCEMENT
OF SCIENCE
1333 H Street NW
Washington, DC 20005
(202-326-6400)

INTERNATIONAL, NATIONAL, AND REGIONAL ORGANIZATIONS

International

ACID RAIN FOUNDATION, INC.
FOR ENVIRONMENTAL TECHNOLOGIES
P.O. Box 473662
Aurora, CO 80047
(303-364-2904)

ASSOCIATION FOR THE ENVIRONMENTAL HEALTH
OF SOILS
150 Fearing Street
Amherst, MA 01002
(413-549-5170)

CENTER FOR ENVIRONMENTAL INFORMATION, INC.
50 West Main Street
Rochester, NY 14614
(716-262-2870)

ENVIRONMENTAL DEFENSE FUND
257 Park Avenue South
New York, NY 10010
(212-505-2100)

ENTOMOLOGICAL SOCIETY OF AMERICA
9301 Annapolis Road
Lanham, MD 20706-3115
(301-731-4535)

FOUNDATION FOR ENVIRONMENTAL EDUCATION
Society of Environmental Toxicology and Chemistry
1010 North 12th Avenue
Pensacola, FL 35201
(904-469-1500)

FRESHWATER FOUNDATION
Spring Hill Center
725 County Road 6
Wayzata, MN 55391
(612-449-0092)

INTERNATIONAL CLEARINGHOUSE FOR ENVIRONMENTAL TECHNOLOGIES
P.O. Box 473663
Aurora, CO 80047
(303-364-2904)

INTERNATIONAL COUNCIL FOR ENVIRONMENTAL LAW
Adenaueralle 214
D-5300 Bonn 1
Federal Republic of Germany
(49-228-2692-240)

INTERNATIONAL ECOLOGY SOCIETY
1471 Barclay Street
St.Paul, MN 55106-1405
(612-774-4971)

INTERNATIONAL SOCIETY OF CHEMICAL ECOLOGY
University of South Florida
Department of Biology
4202 Fowler Avenue
Tampa, FL 33620
(813-974-2336)

NATURAL RESOURCES DEFENSE COUNCIL (NRDC)
40 West 20th Street
New York, NY 10011
(212-727-2700)

SIERRA CLUB
730 Polk Street
San Francisco, CA 94109
(415-776-2211)

SOCIETY OF ENVIRONMENTAL TOXICOLOGY
AND CHEMISTRY
1010 North 12th Avenue
Pensacola, FL 32501-3307
(904-469-1500)

UNITED NATIONS ATOMIC ENERGY AGENCY
2 U.N. Plaza
New York, NY 10017
(212-963-6011)

UNITED NATIONS ENVIRONMENT PROGRAMME
(UNEP)
2 U.N. Plaza, Room 803
New York, NY 10017
(212-963-8138)

UNITED NATIONS FOOD AND AGRICULTURE
ORGANIZATION
2 U.N. Plaza
New York, NY 10017
(212-963-6039)

UNITED NATIONS SCIENTIFIC AND CULTURAL
ORGANIZATION
2 U.N. Plaza
New York, NY 10017

WASHINGTON WORKSHOPS FOUNDATION
Global Environment Seminar
3222 N Street NW, Suite 340
Washington, DC 20007
(202-965-3434)

WILDLIFE SOCIETY
5410 Grosvenor Lane
Bethesda, MD 20814
(301-897-9770)

WORLD ENVIRONMENT CENTER
419 Park Avenue South, Suite 1800
New York, NY 10016
(212-683-4700)

WORLD RESEARCH FOUNDATION
15300 Ventura Boulevard, Suite 405
Sherman Oaks, CA 91403
(818-907-5483)

WORLD RESOURCES INSTITUTE
1709 New York Avenue NW, Suite 700
Washington, DC 20006
(202-638-6300)

WORLD WILDLIFE FUND
1250 24th Street NW
Washington, DC 20037
(202-293-4800)

WORLDWATCH INSTITUTE
1776 Massachusetts Avenue NW
Washington, DC 20036
(202-452-1999)

National

AGRICULTURE RESEARCH INSTITUTE
9650 Rockville Pike
Bethesda, MD 20814
(301-530-7122)

AIR AND WASTE MANAGEMENT ASSOCIATION
1 Gateway Center, 3rd Floor
Pittsburgh, PA 15222
(412-232-3444)

ALLIANCE FOR ENVIRONMENTAL EDUCATION,
INC.
9309 Center Street, Suite 101
Manassas, VA 22110
(703-330-5667)

AMERICAN SOCIETY OF AGRONOMY
677 South Segoe Road
Madison, WI 53711
(608-273-8080)

AMERICAN WATER RESOURCES ASSOCIATION
950 Herndon Parkway, Suite 300
Herndon, VA 22070
(703-904-1225)

CENTER FOR ENVIRONMENTAL MANAGEMENT
INFORMATION
P.O. Box 23769
Washington, DC 20026-3769
(800-736-3282)

CENTER FOR ENVIRONMENTAL PHYSIOLOGY
5632 Connecticut Avenue
P.O. Box 6359
Washington, DC 20015
(202-363-9575)

CENTERS FOR DISEASE CONTROL
1600 Clifton Road
Atlanta, GA 30333
(404-639-3311)

CHEM TREC CENTER—NON-EMERGENCY SERVICES
2501 M Street NW
Washington, DC 20008
(800-CMA-8200)

CITIZENS CLEARINGHOUSE FOR HAZARDOUS
WASTE
P.O. Box 6806
Falls Church, VA 22040

CRAIGHEAD ENVIRONMENTAL RESEARCH
INSTITUTE
Box 156
Moose, WY 83012
(307-733-3387)

ECOLOGICAL SOCIETY OF AMERICA
2010 Massachusetts Avenue NW
Washington, DC 20036
(202-833-8773)

ENVIRONMENTAL LAW INSTITUTE
1616 P Street NW, Suite 200
Washington, DC 20036
(202-328-5150)

ENVIRONMENTAL STUDIES INSTITUTE
800 Garden Street, Suite D
Santa Barbara, CA 93101
(805-965-5010)

ERIC CLEARINGHOUSE FOR SCIENCE,
MATHEMATICS AND ENVIRONMENTAL EDUCATION
1929 Kenny Road
Columbus, OH 43210-1080
(800-276-0462)

GOVERNMENT INSTITUTES, INC.
4 Research Place, Suite 200
Rockville, MD 20850
(301-921-2300)

GREEN SEAL
1730 Rhode Island Avenue NW, Suite 1050
Washington, DC 20036-3101
(202-331-7337)

HAZARDOUS MATERIALS CONTROL RESEARCH
INSTITUTE
1 Church Street, Suite 200
Rockville, MD 20850-4129
(301-251-1900)

NATIONAL AERONAUTICS & SPACE
ADMINISTRATION (NASA)
GODDARD SPACE FLIGHT CENTER
2880 Broadway
New York, NY 10025
(212-678-5500)

NATIONAL AIR TOXICS INFORMATION
CLEARINGHOUSE (NATICH)
U.S. Environmental Protection Agency
Office of Air Quality Planning and Standards, MD-13
Research Triangle Park, NC 27711
(919-541 0850)

NATIONAL AUDUBON SOCIETY
700 Broadway
New York, NY 10003
(212-979-3000)

NATIONAL ENVIRONMENTAL HEALTH
ASSOCIATION
720 South Colorado Boulevard
South Tower, Suite 970
Denver, CO 80222
(303-756-9090)

NATIONAL INSTITUTES OF HEALTH
or
NATIONAL CANCER INSTITUTE
9000 Rockville Pike
Bethesda, MD 20892-0001

NATIONAL WATER RESOURCES ASSOCIATION
3800 North Fairfax Drive, Suite 4
Arlington, VA 22203
(703-524-1544)

NATIONAL WILDLIFE FEDERATION
1400 16th Street NW
Washington, DC 20036-2266
(202-797-6800)

PUBLIC LANDS FOUNDATION
P.O. Box 10403
McLean, VA 22102
(703-790-1988)

RENEW AMERICA
1400 16th Street NW, Suite 710
Washington, DC 20036
(202-232-2252)

SMITHSONIAN ASTROPHYSICAL OBSERVATORY
(SAO)
60 Garden Street
Cambridge, MA 02138
(617-495-7461)

SOIL & WATER CONSERVATION SOCIETY
7515 Northeast Ankeny Road
Ankeny, IA 50021
(515-289-2331)

WOMEN'S ENVIRONMENT AND DEVELOPMENT
ORGANIZATION
Women USA Fund, Inc.
845 Third Avenue 15th Floor
New York, NY 10021
(212-759-7982)

Regional

ACADEMY OF NATURAL SCIENCES
OF PHILADELPHIA
ENVIRONMENTAL RESEARCH DIVISION
1900 Benjamin Franklin Parkway
Philadelphia, PA 19103
(215-299-1110)

AMERICAN HEALTH FOUNDATION
320 East 42nd Street
New York, NY 10017
(212-953-1900)

ARIZONA STATE UNIVERSITY CENTER FOR
ENVIRONMENTAL STUDIES
Tempe, AZ 85287-3211
(602-965-2975)

ASSOCIATED UNIVERSITIES FOR TOXICOLOGY
RESEARCH AND EDUCATION (AUTRE)
4301 W. Markham
UAMS 522-7
Little Rock, AR 72205-7122
(501-686-6501)

AUBURN UNIVERSITY WATER RESOURCES
RESEARCH INSTITUTE (WRRI)
202 Hargis Hall
Auburn University, AL 36849
(205-844-5080)

BAYLOR COLLEGE OF MEDICINE BIRTH DEFECTS
CENTER
6621 Fannen Street
Houston, TX 77030
(713-791-3261)

BROWN UNIVERSITY ANIMAL CARE FACILITY
Box C
Providence, RI 02912
(401-863-3223)

CARL HAYDEN BEE RESEARCH CENTER
2000 East Allen Road
Tucson, AZ 86719
(602-629-6380)

CASE WESTERN RESERVE UNIVERSITY
DEVELOPMENTAL BIOLOGY CENTER (DBC)
Cleveland, OH 44106
(216-368-3430)

CENTER FOR BIOENVIRONMENTAL RESEARCH
2800 Victory Boulevard
Staten Island, NY 10314
(718-982-3921)

CENTER OF ALLERGY & IMMUNOLOGICAL
DISORDERS
1200 Moursund Avenue
Houston, TX 77030
(713-791-4219)

CENTER FOR NATURAL RESOURCES
3123 McCarty Hall
Gainesville, FL 32611
(904-392-7622)

CITY COLLEGE OF THE CITY UNIVERSITY
OF NEW YORK
URBAN MARINE AND FRESHWATER INSTITUTE
Office of Academic Affairs
535 East 80th Street
New York, NY 10021
(212-794-5681)

CLEMSON UNIVERSITY
PEE DEE RESEARCH & EDUCATION CENTER
FOR AGRICULTURE
P.O. Box 271
Florence, SC 29503
(803-662-3526)

COLUMBIA UNIVERSITY
CORNELL PLANTATIONS
1 Plantation Road
Ithaca, NY 14850
(607-256-3020)

DUKE UNIVERSITY MARINE BIOMEDICAL CENTER
Duke University
Beaufort, NC 28516
(919-728-2111)

GEORGIA CROP REPORTING SERVICE
Federal Building, Suite 320
Athens, GA 30613
(404-546-2236)

HARVARD UNIVERSITY
KRESGE CENTER FOR ENVIRONMENTAL HEALTH
665 Huntington Avenue
Boston, MA 62115
(617-732-1272)

MCDONNELL CENTER FOR SPACE SCIENCES
Box 1105
St. Louis, MO 63130
(314-889-6225)

MEMORIAL SLOAN-KETTERING INSTITUTE
FOR CANCER RESEARCH
1275 York Avenue
New York, NY 10021
(800-422-6237)

NEBRASKA STATEWIDE ARBORETUM
112 Forestry Sciences Laboratory
Lincoln, NE 68583
(402-472-2971)

NEW ENGLAND PLANT, SOIL & WATER
LABORATORY (NEPSWL)
University of Maine at Orono
Orono, ME 04469
(207-581-2216)

NEW YORK ACADEMY OF SCIENCES
2 East 63rd Street
New York, NY 10021
(212-838-0230)

NEW YORK BOTANICAL GARDEN
Southern Boulevard and 200th Street
Bronx, NY 10458
(718-220-8700)

PURDUE UNIVERSITY
SOUTHERN INDIANA–PURDUE AGRICULTURE
CENTER
R.R. 2
Dubois, IN 47527
(812-678-3410)

STATE UNIVERSITY OF NEW YORK
COLLEGE OF ENVIRONMENTAL SCIENCE
AND FORESTRY
INSTITUTE FOR ENVIRONMENTAL POLICY
AND PLANNING
Bray Hall, Room 320
Syracuse, NY 13210
(315-470-6636)

THORNE ECOLOGICAL INSTITUTE
5398 Manhattan Circle
Boulder, CO 80303
(303-499-3647)

UNIVERSITY OF NORTH CAROLINA
AT CHAPEL HILL
INSTITUTE FOR ENVIRONMENTAL STUDIES
CB 7410
315 Pittsboro Street
Chapel Hill, NC 27599-7410
(919-962-2211)

GOVERNMENT AGENCIES

BUREAU OF OCEANS & INTERNATIONAL
ENVIRONMENTAL & SCIENTIFIC AFFAIRS
2201 C Street NW
Washington, DC 20520
(202-647-1554)

FEDERAL MARITIME COMMISSION
6 World Trade Center
New York, NY 10047
(212-264-1425)

NATIONAL OCEANIC & ATMOSPHERIC
ADMINISTRATION
Office of Public Affairs, Room 6013
14th Street and Constitution Avenue NW
Washington, DC 20230
(202-482-6090)

UNITED STATES DEPARTMENT OF AGRICULTURE
(USDA)
ARS VISITORS CENTER
Building 302, BARC-East
Powdermill Road
Beltsville, MD 20705
(301-344-2403)

UNITED STATES DEPARTMENT OF AGRICULTURE
(USDA)
U.S. FOREST SERVICE
Information Office
201 14th Street SW
Washington, DC 20090
(202-205-1760)

UNITED STATES DEPARTMENT OF JUSTICE
DRUG ENFORCEMENT ADMINISTRATION
600–700 Army Navy Drive
Arlington, VA 20202
(202-307-1000)

UNITED STATES ENVIRONMENTAL PROTECTION
AGENCY (USEPA)
401 M Street SW
Washington, DC 20460
(202-260-2090)

UNITED STATES FISH AND WILDLIFE SERVICE
1849 C Street NW
Washington, DC 20240
(202-208-5634)

SCIENTIFIC SUPPLY COMPANIES

BAXTER SCIENTIFIC PRODUCTS
1430 Waukegan Road
McGraw Park, IL 60085-6787

FISHER SCIENTIFIC
711 Forbes Avenue
Pittsburgh, PA 15219
(Sold by Fisher:
Bay Water Technology Manual
Standard Methods Manual for Water Analysis
EPA-approved analytical methods)

SARGENT-WELCH SCIENTIFIC COMPANY
7300 North Linden Avenue
P.O. Box 1026
Skokie, IL 60077

THOMAS SCIENTIFIC APPARATUS AND REAGENTS
99 High Hill Road at I-295
P.O. Box 99
Swedesboro, NJ 08085-0099

VWR DIRECT
1430 Waukegan Road
McGraw Park, IL 60085

DATABASES

AEROMETRIC INFORMATION RETRIEVAL SYSTEM
(AIRS)
U.S. Environmental Protection Agency
Office of Air and Radiation
Office of Air Quality Planning and Standards
Information Management Group
Research Triangle Park, NC 27711
(919-541-5454)

AMBIENT WATER QUALITY CRITERIA DOCUMENTS
U.S. Department of Commerce
National Technical Information Service
Springfield, VA 22161
(703-487-4650)

BIOSIS PREVIEWS
BIOSIS
2100 Arch Street
Philadelphia, PA 19103-1399
(800-523-4806)

CHEMEST
Technical Database Services, Inc.
135 West 50th Street
New York, NY 10020-1201
(212-245-0044)

EARTH SCIENCE DATA DIRECTORY
U.S. Geological Survey
814 National Center
Reston, VA 22092
(703-648-7182)

LEXIS-NEXIS
9443 Springboro Pike
P.O. Box 933
Dayton, OH 45401-9964
(800-227-4908)

GENERAL WEB SITES ON THE ENVIRONMENT

U.S. Environmental Protection Agency home page (EPA)
http://www.epa.gov

National Environmental Information Resources Center
http://gwis.circ.gwu.edu:80/~greenu

List of State Environmental Agencies
http://www.tribnet.com/environ/emv_stat.htm

List of University Environmental Web Sites
http://bigmac.civil.mtu.edu/aeep/univ.html

Environmental Organization Webdirectory
http://www.webdirectory.com/

Electronic Green Journal
http://gopher.uidaho.edu:70/UI_gopher/library/egj

EnviroWeb
http://www.enviorlink.org

National Audubon Society
http://www.audubon.org/audubon/